HULA

A Boat's Story

Emy Thomas

Cover painting by Laurie Ingersoll

Maps by Cynthia Hatfield

ISBN: 9798850061012

Also by Emy Thomas

Non-Fiction

Home Is Where the Boat Is

Life in the Left Lane

A Most Unlikely Story

(A Short Memoir of a Long Life)

North Atlantic
Bermuda
Gulf of Mexico
Bahamas
Cuba
Yucatan
Dominican Republic
Virgin Islands
Jamaica
Puerto Rico
St. Martin
Antigua
Guadeloupe
Belize
Caribbean Sea
Barbados
Grenada
Curacao
Venezuela
Trinidad
Panama
Cocos
equator

Dedication

To Peter and *Solanderi*

Russia
Aleutions
North Pacific
Victoria
Japan
U.S.A.
Santa Barbara
China
Hawaii
Mexico
Taiwan
Guatemala
Philippines
Guam
Cocos
Truk
equator
Galapagos
Equador
New Guinea
Solomon
Marquesas
Samoa
Fiji
Tahiti
Tonga
Australia
South Pacific
Tasmania
New Zealand

Acknowledgements

Thanks to Apple Gidley for positive support throughout various versions of this story. And to my other readers Tom Thomas, Laurie Ingersoll, Rich Difede and especially Shirley Ziegler, who also became my publishing facilitator, a big job. Thanks to Laurie Ingersoll for her patience, talent and skill producing multiple versions of the cover art and to Michael Crumpton for the cover layout. To Cynthia Hatfield for the maps, and to Colleen Sullivan for her kind patience with technical assistance.

Hula
A Boat's Story

The Launching

She floats! they shouted as I plopped into the bay.

As my hulls settled into the gentle salt waters of the Caribbean Sea I felt a sensuous bliss. This is where I belong. I am a boat and I float.

A few of the onlookers were obviously relieved, but all of the small crowd gathered at the dock was just genuinely happy that Jack had actually pulled it off. After 18 months of hard labor, he had launched a boat—me.

It's no surprise to me. Even when I was just sheets of plywood and rolls of fiberglass I absorbed his mission and helped in every way I could. His dream soon became mine and we worked together like a well-synchronized machine.

Now, after hours of being jostled around, then rolled into the sea on a series of logs, I was floating. What a fabulous feeling that is!

Jack and his pals grabbed the lines attached to me and pulled me to a rickety dock that will be my home while Jack finishes me. They swarmed onto my decks and the beer began to flow. Most of them are familiar to me as they have often patted my sides when they came to our yard to lend a hand or just schmooze away the evening.

Here is the invitation Jack wrote for my party:

LAUNCHING BASH
Jack's Catamaran
Date: March 10, 1980
Time: 2 PM til...
Place: On Deck by the Dock
at The Yard at Secret Harbor
St. Croix, Virgin Islands

Before the group got too out of hand Jack went up to my port bow, raised a bottle of Cruzan Rum until he had everyone's attention, then announced I christen you *Hula Hula* as he smashed the bottle on my bow.

Much to my surprise this kinda quiet guy made a nice little speech: Welcome aboard *Hula Hula*. You are now aboard a catamaran that was designed by a Welshman and built by me, a New Englander, to sail around the world. I have named her *Hula Hu*la because her design was inspired by the ancient Hawaiian double canoes, the first vessels ever to sail long distances in open seas. And because it seems appropriate for a double canoe to have a double name and a Hawaiian one at that. And because I hope our lives together—mine and *Hula's*—will be a dance.

That last line really got me. I had no idea my creator could be so sweet! Or eloquent.

He put on some Hawaiian music, took off his shirt, tied a sarong around his hips and distributed a bagful of paper leis to the celebrants and the party took off.

I basked in the attention and the appreciation I seemed to acquire that day. I am no longer just a construction site. I am a floating boat and quite a special one.

I'm a Wharram Cat and I'm going to sail around the world! My sister catamarans are known to travel successfully all over the globe and I can't wait to meet some of them. We are not known for our beauty or our speed. We will never win a race but we will get you to where you want to go!

I'm getting way ahead of myself because there's still a lot of work to do but you can imagine how exciting it is to be actually floating.

A lot of my partyers are kinda floating too.

No! They're trying to dance the Hula Hula! He's got the right music on but these young people, most of them Caribbean islanders or ex-pats from the States, are moving like it's calypso! Oh well!

I'm a good party boat already. Unlike most yachts, I have a nice big deck between my two hulls so there's plenty of room to do whatever they want with their feet and hips.

Yes I can call myself a yacht. I know it sounds snooty but in the nautical world it refers to all pleasure boats.

I should let you know right away that there will be a lot of vocabulary in this story you might not understand. That's because English-speaking sea-farers from way back have been naming the things they work with and no one has ever dared change a word. There's a special glossary at the end of the book devoted to this secret language, which Jack calls Nauticalese, so whenever you run across a mystery word please check there.

I'm hoping this party won't go on all night. I've had a really big day and Jack has brought his air mattress from the shed he's been sleeping in to his bunk-to-be. We're going to sleep together tonight for the first time.

***

The seed of my birth was planted four years ago, in 1976, when Jack watched a new PBS documentary called *The Navigators*. It follows *Hokule'a*, a 60-foot modern reconstruction of the double-hulled Hawaiian voyaging canoes that sailed the Pacific centuries ago, when she made her maiden voyage of 2,400 miles from Hawaii to Tahiti earlier that year.

The navigator was one of a handful of men left in the Pacific if not the world who can navigate without instruments. He guided *Hokule'a* over 2,400 miles, on a route he had never sailed before, using only his knowledge of the sun, stars, swells, waves, currents, birds and fish to point his way.

Jack was blown away by the film and it changed his life. Oh don't worry. He will not be navigating without instruments, but he will be sailing an updated replica of that very same double-hulled canoe.

The Navigators so enthralled him he started researching and soon discovered the Wharram cats and their do-it-yourself plans for boats that were based on those Hawaiian voyagers! In all sizes that were remarkably sea-worthy and were enjoying cruising in all corners of the globe!

He immediately ordered plans for an Oro, a 45-footer, and four years later the result is me!

A Little Backstory

Jack grew up living on the island of Martha's Vineyard in Massachusetts and sailing off the New England coast on his father's 40-foot ketch. Most of his life he knew he would have a boat of his own as soon as he could afford it and eventually take off for other parts of the world.

He wants to go slowly, to live in other places, not just visit them. He intends to work his way around the world and he is well-prepared for that. His dad is a jack of all trades who owns a hardware store. He made sure his son grew up equally skilled.

But he also insisted Jack go to college. Jack got a degree in engineering to please his parents but he knew all along that he was happier using his hands than his brain. He never wanted an office job, or anything 9-5. His parents were disappointed but not surprised when after graduation he rented an abandoned shop in Vineyard Haven and started a carpentry business.

He had worked most summers since he was 16 for the best finish carpenter on the island and had his blessing. His old boss sent a lot of work his way, and Jack attracted the younger and less affluent residents.

Among other things he made simple but

handsome furniture for the homes there, still mostly nicely weathered shingle cottages.

He's quite creative and began carving his wood scraps in slow periods. When a kooky potential client noticed some elegant small bowls and miniature canoes lying around the shop he commissioned Jack to make him a table fashioned after his dog!

Jack created a coffee table that was more or less a wood sculpture of the client's Golden Retriever. It was made from a white oak log, stained gold. The client was so thrilled he sent a photo of it to the *Vineyard Weekly*, which printed it, and Jack was never without enough clients from then on. Both regular and artistic commissions.

In good weather he still sailed on the family boat on weekends, or joined friends on their smaller boats, but when all those vessels were stowed for the winter, he felt deprived.

I was just a dream then, in the mid 70s—a few scribbles on scraps of paper in Jack's carpentry shop.

But one nasty February he closed shop for two weeks and flew to the island of St. Croix to see a college friend who had moved there. The friend, Ben, who also had an engineering degree, was making a living waiting on tables at night in an open-air gourmet restaurant. They spent almost every day in or on the water, snorkeling, scuba diving, wind-surfing or sailing a small boat Ben had acquired when the yacht club discarded it as too beat up for its sailing classes.

Back on the Vineyard Jack dreamed about that vacation day and night. Serendipitously, he happened to catch the airing of *The Navigators* then and soon he was hooked on the idea of having a double-hulled boat of his own.

The few catamarans he had seen were not at all suitable for crossing oceans but with just a little research Jack soon found the Wharram do-it-yourself plans for cruising boats based on those very same ancient Polynesian sailing double canoes in the documentary! He immediately ordered plans for a 45-footer.

He had never sailed on a multihull and he had never worked with fiberglass before but these boats were cheap and known to successfully cruise the world and since that's what he was going to do suddenly everything meshed. He had a plan. He would build the boat on St. Croix.

On his 30th birthday he announced to his family, friends and clients that he was moving to St. Croix in the Virgin Islands. Thanks to Ben he had a cheap place to live.

He didn't mention to the Vineyard crowd that he had saved enough money to support himself for a couple of years and he planned to build a boat! He didn't mention to these Yankees who thought that Sunday racing in their sleek monohulls was the best of the sailing life that his boat would be a catamaran that he would build himself and would sail around the world.

Secret Harbor

Jack's friend Ben rents a big old house on the waterfront in Secret Harbor with an often-changing cast of roommates. During Jack's visit to St. Croix, he had nosed around the big property and found an abandoned roofless shed that he realized he could fix up as a workshop. The yard was big enough to accommodate two 45-foot hulls without taking over the area where the roommates had a makeshift patio and barbecue. The house had been built by a fisherman years before so already had the remains of a dock he could fix up easily. It would be the perfect set-up for him.

He presented his plan to Ben, who shares the house with two other guys. They're all waiters and work at different restaurants in town. Since there are only three bedrooms they agreed that Jack could share the rest of the house and plug into their power but he would have to sleep in the shed. No problem. He would pay a quarter share of the rent and utilities, saving them each a nice chunk of cash.

As soon as he arrived on the island Jack bought a tough old black pickup with one blue door and one yellow, loaded it with plywood, slapped a roof on his shed, cleaned it and repaired the wood floor, ran a series of extension cords from the house, bought an air mattress and some pillows and moved in. A box containing all his carpentry tools arrived and he was good to go.

The Creation

I can't tell you much about the actual building because, of course, I wasn't around then. I know only that my two plywood hulls were built upside down on the level part of the lawn. Those two huge canoes were then turned right side up for the addition of decks and in the middle of each a cabin with standing room. All the plywood was covered with fiberglass.

There are several openings to let air in. Horizontal hatches that open from cabin tops on hinges and perpendicular portholes on the sides that have sliding windows.

The two hulls are connected by four crossbeams. The two in the center support a wood deck between the cabins. Eventually there will be netting between the hulls fore and aft of that deck.

Those crossbeams, Jack tells the boys, have rubber mounts that make all the difference between me and other catamarans. They give Wharrams the ability to adjust to the ocean's movement, to go with the flow, to sail around the world!

The yard where this was all created and put together now has a clear path from the shed where Jack keeps his tools to the dock where I am being finished. He's back and forth all day. A talented carpenter, he is transforming me into a seaworthy yacht and also an attractive home.

He is a master at making something out of nothing and most of his materials were salvaged or hand-me-downed.

Half of my portholes are quite classy. They were salvaged from an elegant motor yacht that got wrecked in one of the hurricanes. These will go in the cabin tops. The other half of my portholes, those that are already installed over the bunks, are not quite so pretty. They are also salvaged.

Most important, he already has two masts! They are lying in the yard until all the other work is done. They had been stored at the boatyard in town since a cruising yacht tried to enter Christiansted harbor, a tricky channel, at night and instead crashed onto Long Reef. The crew was rescued but the boat was trashed in no time. There wasn't much to salvage but the masts were basically intact so have been waiting for a taker—us! They are exactly the right size for the rig Jack has planned.

Wharram's plans for the Oro actually call for a lateen rig. Jack mentioned this to his roommates one afternoon over beers. Noone can imagine why the designer thought that ancient Mediterranean-style sail would be fitting for his Polynesian-style boats.

Jack plans to make my rig that of the standard Western ketch—a tall mainmast forward (toward the front of the boat) and a shorter one, a mizzen mast, aft (toward the rear). He's learned that most other Wharram builders have done that and found it worked well.

A Visitor

Hello?

Jack didn't hear the call because he was deep in my starboard hull finishing a storage area. But I perked up at once. A girl! And she's looking for Jack! She came onto our dock gingerly, touched my side gently, and said You must be the one. I was smitten, of course. Most girls don't pay much attention to a DIY like me. She called again and Jack finally emerged on deck, all sweaty, dirty and smelly.

Hi she said. You must be Jack. I'm a reporter for *The Island Times* and we'd like to do a story on you. And this boat. You built it all yourself? You're planning to sail it around the world? Jack managed a couple of nods, then finally came to and said Okay. Come aboard. Oh I'm sorry, there's no place to sit yet. Do you mind sitting on the cabin top? Have some water. He poured into two cups from his cooler, clinked hers with his and said Welcome aboard.

She asked for a tour of the boat. He said You'll have to use your imagination a lot and preceded her down the port companionway. This will be the galley he said. That's boat talk for kitchen. All she could see was a small sink and a pump.

How will you get water she asked.

See this pump he answered. I get salt water up straight from the sea. I hope to be in clean water most of the time and will use mostly salt water.

That's it? she asked incredulously. Oh, I'll have some fresh water too he smiled. Once I make my awning I'll be able to collect my own rain water.

Oh, how do you do that?

I'll cut two holes in the awning, one over each hull, and attach plastic tubes that will carry the water into jugs. The awning will always be up when I'm at anchor.

Oh, clever! How will you cook? she asked.

I'll get some kind of stove, probably one or two burners that will run on propane. Food will be stored up here he said indicating the space under the forward bunk. You have to carry a lot of cans and stuff when you're going long distances.

What about refrigeration?

What's that? he replied.

My gosh, he's teasing her! And I think she's enjoying it!

No, look, he explained. This is going to be a very basic boat. You've heard of the KISS principle? You know what that is?

Oh, you mean Keep It Simple Stupid?

That one.

But how far can you go with that?

All the way he replied. I won't have electricity. I won't have an engine.

You're kidding me of course.

No. Why would I do a thing like that?

Because you think I'm gullible?

I'm not kidding he replied. Instead of a diesel engine I'll just have the little outboard I now have on my dingy and have kerosene lamps instead of electric lights. She thought about that for a while. You're either a very good sailor, she decided, or a total nut case. That seems very dangerous.

Well he replied, I hope I can sail as well as all those guys in the last centuries who never had engines. I'll be living like them anyway--no electricity, no refrigeration, no running water. You wanna see the rest of the boat?

She nodded yes. Up one ladder, across the deck and down another ladder to the starboard hull, where two bunks were in the process of being built in, one forward and one aft. In the middle was a wider area with standing room where, Jack explained, there would be cupboards and lockers for clothes, a little basin and a head.

A head? she asked looking quite perplexed. Oh wow, Jack smiled. You don't know what a head is! That's the john, the toilet! It takes a bit to get used to. You can't just push a handle and expect everything to disappear. You have to pump. Don't look so horrified. Once you get used to it there's nothing to it.

Back on deck she said there are so many more questions! We're standing on the deck, right? How will you keep from falling overboard?

Ohmygod he laughed. There's so much more to explain. I'll be making bulwarks. Those are plywood extensions above the hulls. They come to about here he said tapping his knee. And above them will be lifelines—plastic lines looped through stainless steel stanchions that will be secured in the hulls. They're up to about here he said this time tapping his hip. You can grab ahold of them any time you're near the edge.

That sounds good she said. And this is a sailboat, right? Where will the sail be?

Well there'll be a few sails he said. See those two long poles lying on the grass over there? They'll be my two masts. The longer one goes here he said walking forward and the shorter one here he said walking aft. There's actually a good story about them. Were you in St. Croix when that yacht went aground on Long Reef a year or so ago? Well those masts are about all that's left of that boat. It so happens they are the perfect size for mine!

What luck she said. It's none of my business but I get the impression you're doing this on quite a limited budget so I would guess that was quite a welcome find.

You're not kidding, Jack said. Yeah. I'm an accomplished scrounge. See these porthole windows? They're salvaged from a yacht that was wrecked. And that mahogany table below in the salon? I found it at a dumpster. A couple of the legs were broken. The top was all I needed.

Salon? she questioned. Oh yeah he laughed. That's the snooty sounding word for the part of the cabin that serves as a living/dining area on a traditional boat. The Brits call it a saloon! Mine is just a table and bench tucked under the deck but it's still referred to as the salon.

When they finished the tour Jack said Well I guess that's it.

Thanks she said. Now just a few personal questions, okay?

Jack's never been interviewed before so was a bit reticent to part with any information. So just about the only thing she learned about him as a person is I'm from Massachusetts—Martha's Vineyard—and I worked as a carpenter there. Yeah I went to college, I have an engineering degree. Yeah my dad has a sailboat so I grew up sailing off the East Coast. Oh yeah I'm doing this alone. I'll probably pick up crew here and there along the way but I'll be working my way around the world so there will be a lot of stop and go as I pick up jobs.

When she asked What brought you to St. Croix? he mumbled It seemed like a good place to build a boat. He's not exactly eloquent!

But the girl seemed impressed, and I think she thinks Jack is pretty cute. So I'm excited! He hasn't paid much attention to girls since I've been around. He really needs a mate. That's nautical for assistant but it's also English for a loving partner. I'm hoping for someone who fits both job descriptions.

She seemed happy enough with the encounter and when she was ready to go jumped onto the dock quite gracefully—I was happy to see that—and they both waved goodbye with big smiles. Jack went back to his sawing and gluing immediately but a few minutes later I heard him mutter I didn't ask her name! DAMN! I second that! He'll never get a girl!

***

A few days later Jack's buddy Jim came racing onto the dock. Waving *The Island Times*. You're on the front page he yelled. Sure enough. Her photos of him at work were pretty good and as he read the article he looked first shocked at her descriptions—she called me bashful and disheveled—then alarmed—amazingly creative and resourceful—and finally pleased. She called me a throwback! A figure from the past. She thinks I'm something out of Joseph Conrad!

Who's he? asked Jim.

My favorite author! replied Jack. And here's her byline! Jill Sanford.

Then he did a double-take. What's wrong? asked Jim. Jack revealed Oh, my mother was always looking for a Jill to be my wife. He looked stricken. Jim went HUH? Jack said You know: Jack and Jill, the nursery rhyme? My mother thinks it would be cute for me to marry a Jill! Oh yeah Jim laughed. And you don't wanna marry this girl? I'm not gonna marry any girl Jack said.

I'm going to sail around the world—remember? Oh yeah. Right Jim said. I guess you don't have to worry about marrying. I don't think there are any chicks around who would wanna join you.

Well, this was a new subject for me. I'd never really thought about who might be going to sea with us, if anyone. I've always thought of just me and Jack. But I gotta say I felt a little thrill just hearing that word. Wife. And now that I think about it I really would like him to have a companion. I don't want to go to sea with a lonely bachelor.

***

Jack and Jill did start to spend some time together and I could tell they liked each other a lot. They played up the brother and sister bit from the nursery rhyme hoping others would think of them as just friends and not assume they were a couple. He didn't intend to commit to anything but his boat and his freedom and she certainly wasn't going to get too involved with a guy who lived in a shed and lavished all his time and money on—I must admit—a not very beautiful boat.

The first time they joined the roommates and friends for a Sunday afternoon barbecue Jack made a point of introducing her as a Friend. Also, since almost everyone has heard only the first verse of the nursery rhyme, which leaves you assuming Jack dies, he insisted they hear the whole thing so they too would know the ending.

When it looked like everyone who was expected had arrived Jack called for attention and announced I'd like you all to meet my friend Jill and to apologize for introducing a nursery rhyme into our yard. But since I know there will be a lot of kidding around about our names I want to make sure you've got the whole story. I just looked up the nursery rhyme myself and was surprised and relieved to find there are three verses and Jack survives and that he and Jill were siblings. I think it will be helpful for all of you to know what I know. So here goes:

Jack and Jill went up the hill
to fetch a pail of water.
Jack fell down and broke his crown
And Jill came tumbling after.

Then up got Jack and said to Jill
As in his arms he took her
Brush off that dirt for you're not hurt,
Let's fetch that pail of water.

So Jack and Jill went up the hill

To fetch the pail of water,
And took it home to Mother dear
Who thanked her son and daughter.

The reading was greeted with hooting and applause and laughter and some literary commentary:

That's a pretty stupid poem isn't it!

No wonder we never heard the other verses!

How did that ever get to be so popular!

I hope you don't go around performing that!

Is this an announcement of some kind?

Jack jumped on that last comment and said no, we are not siblings but I don't have a sister and she doesn't have a brother so maybe that's why we're friends.

Ben came to the rescue and said thanks for giving us the whole story. I never knew anything but the first stanza and always assumed Jack had died! I'm so glad to know he recovered! Now let's eat!

Back to Work

It took many more months to put me together just enough to go for a trial run. First Jack made my bulwarks, which seemed to take forever but I gotta say make me look a whole lot better. I am no longer just two hulls with center cabins poking through.

The bulwarks are basically the on-deck "walls". Ours are wood strips rising from the hulls' edges all around except where the deck joins the center cabins. Jack has shaped ours with a little swoop fore and aft that make me look almost streamlined! And they make it clear that I will indeed be going to sea! A bulwark's primary purpose is to keep the people aboard aboard!

Then came the stanchions and lifelines. Stanchions are stainless steel poles attached to the deck inside the bulwarks. They rise above the bulwarks to about hip height. They have loops at the top through which vinyl-coated wire ropes are pulled all around the boat. These give you something to grab onto when you're moving about near the edge of the deck. We have two rows of landlines! Jack currently hangs his laundry on them.

Those jobs took months, but it wasn't boring. The roommates are fun and help whenever needed and what I really like is that Jill is around more and more. Unless there's something major going on at night that she has to cover for the paper she usually stops by at least for a beer. Sometimes she brings a take-out supper for them and after Jack installed a hotplate she's actually cooked supper a few times!

She warns Don't expect much! I'm not a cook! Yeah, I feed myself but I'm not used to cooking for anyone else. That could be a big problem cuz Jack is a pretty big eater. But he's used to feeding himself too so they muddle through.

The really good news is that Jill now sometimes stays overnight! Oh my gosh! They have the best time! Jill has even introduced sheets! There is still only one makeshift foam mattress aboard but Jill managed to retire the grubby sleeping bag that covered it all this time and introduce actual sheets! The two of them now can't wait to get to bed, which is on deck when the weather is good or below on the bunk when it's not. There's lots of loving for quite a while, then a good sleep.

I love having Jill around and obviously Jack does too. And I think she feels quite at home here now. I hate to think what will happen when I'm finally ready to go.

I think all that's left now are the masts. The things that hold the sails up and make a sailboat a sailboat. All the roommates were aboard to help raise the two masts and secure them in their steps, the metal supports that hold the masts up. Then it was weeks more installing all the wires and lines that connect the masts to me and the sails. Frigging with the rigging Jack calls it.

Months ago he inherited a mainsail from some departing cruisers which he finally cut down and re-sewed on the old industrial sewing machine he bought, he likes to tell everyone, from a bra factory in Puerto Rico when it went out of business. Once he got that rigged he announced Okay! We're going sailing!

The Trial Run

I can't tell you what a thrill it was for me to move for the first time! I really came to life as little wavelets lapped my sides and the breeze filled the mainsail and pushed us gently this way and that.

As we moved from the bay into the actual OCEAN and the waves got bigger and the wind got stronger WOW! What a trip! I have so many parts that have to work together and they all did remarkably well. Jack was all over the place tweaking this and that, while Jill sat at the wheel following Jack's signals—a little more this way, now all the way that way.

They both looked ecstatic and I think we could have sailed forever BUT. We were lacking many essentials. To name just a few of the most important: The rest of the sails (jib and mizzen, at least), anchors and chains.

That little taste of what was waiting for us spurred us on and in no time I was finished—well, several years later and I know I'll never be finished—

but within months I was actually ready to try cruising!

We were going to the British Virgin Islands, which were nearby and hopefully offering jobs. The BVIs are probably the most popular cruising grounds in the Caribbean for U.S. and international yachts, and the new yacht charter business in Tortola, the main island, has taken off. Sailors and non-sailors are coming from all over to charter a boat for a week of prime sailing. If you didn't qualify to skipper a boat a captain was provided for you.

Jack just knew he could make good money there. All those boats coming and going every week needed lots of maintenance, and he was just the guy to do it. He figured if he spent a year there he could make enough of a nest-egg to start his wanderings.

Much to my delight he asked Jill to come with us if she was willing to contribute to the living expenses for that year and she said yes. She couldn't get an official sabbatical from the *Island Times* but they told her it was very likely they'd re-hire her whenever she came back so she sub-let her apartment for a year and moved her yachting wardrobe aboard—a few bikinis, shorts, jeans and T shirts. She had a couple of long silky dresses she wore to parties and they came too, just in case.

The BVIs

Our maiden voyage was a day's sail across the deep, deep Virgin Islands Trench to the British Virgin Islands. Jack called it an uneventful passage, something I subsequently learned was a rare gift, and we had a marvelous time adjusting to the motion, breaking in the rigging, and lapping up the beauty of the sea and sky, the strength of the sun and wind. I was certainly relieved and proud that we didn't have any crises. I like to share the credit for a job well done.

Now we're just leisurely cruising up and down the Sir Francis Drake Channel, a unique body of water protected on both sides from the winds and swells of open seas—the Atlantic to the east and the Caribbean to the west—by a series of lovely little islands, some inhabited, some not.

We find a nice anchorage every night. I'm proud to say that thanks to my shallow draft, we are able to go where most other boats cannot. Three feet of water is all I need under me.

If there's a bar or little resort ashore J&J go in for at least a drink. A few times they've come back with a guy or a couple curious about me. The BVIs are full of cruising yachts from all over the world but catamarans are quite new here and I am probably the first homemade one a lot of these sailors have ever seen!

There are also a lot of charter boats sailing around. Now Jack is sure he can get jobs. So when the owner of the bar/restaurant on Cooper Island told Jack there was a new charter company in Smugglers Cove on Tortola that might be looking for workers we sailed in.

Smugglers Cove

Sure enough my Jack-of-all-trades quickly got a job at the new Antilles Charter company there. He's assisting the head of the shop on the dock where the yachts are quickly made ready after one charter for the next. The boss was delighted to find Jack.

We're all happy! The bay is small and beautiful, with just a few other boats anchored out. I rock gently on my mooring behind a protecting reef. Jack rows his dinghy to his job ashore. And I'm delighted to say Jill usually stays home to work on me.

Oh, one of the first things that happened is that Jill decided she needed her own dinghy. Sure she could take Jack ashore to his job and have the use of his dinghy all day but we saw a new side of Jill when she realized she needed to have her own. She'd never been dependent on anyone before and didn't want to start now.

So J&J went shopping for another dingy on his first day off and scored a real winner. It's a small blue sailing dinghy with a white sail! A family that was leaving the island left it at the local boatyard. Jill paid $100 out of her own pocket. (Otherwise they share expenses. Jill closed her bank account in St. Croix before leaving. There was only a few hundred dollars in it but she immediately put most of it in a jar in the galley. Jack added his entire fortune—another few hundred. They call this pickle jar The Kitty.)

So they can survive for a while and now that Jack is working Jill can stay aboard to work on me but if she wants to she can get in her dinghy and row ashore.

Both dinghies have Hawaiian names of course. Jack's is *Kimo* and hers is *Lilo*. They're my children.

Back to Painting

I'm embarrassed to admit that I looked more like a candidate for the junkyard than a brand new sailing yacht when we arrived here. Once I was seaworthy Jack was so anxious to go sailing he just didn't do anything that wasn't absolutely essential. So now that we're not going anywhere for a while and Jack has a job it makes sense that Jill should concentrate on making me beautiful. Well, at least presentable. We all know I will never win a beauty contest.

I'm finally getting painted! My hulls were given a slap-dash coat of white paint just before I was launched but that was over a year ago so they desperately need a make-over. Jill can't wait.

She is delighted that Jack has given her free rein on colors. She immediately chose turquoise for the hulls and teal for the bulwarks, colors of the Caribbean Sea. Jack is more of a traditionalist and was assuming I'd be white.

Ideally I would be grounded for this paint job because the turquoise will go all the way to the waterline. But Jill doesn't want to be beached any longer than absolutely necessary so has decided she'll paint my sides from her dinghy.

This bay is so placid there is little action along the waterline so she can paint to within a few inches of the line and the next time we beach to clean my bottoms they'll finish painting the sides.

Antilles Charters

J&J's lives here in Smuggler's Cove are centered

around the charter company and its employees. Most of the people on the few other private yachts moored here are also working for AC and the small management crew that came down from the States welcomes everyone in the cove to come ashore at cocktail hour for free drinks!

All the booze that comes off the boats at the end of a charter is deposited on a makeshift bar under a thatched roof, an establishment known as the Bilge Bar. (Bilge, by the way, is the nautical term for the area under the sole (floor boards of a boat) where storm water and other undesirable liquids end up—quite an unappetizing name for a bar I'd say!)

When the AC crew stops work about 4, many of the workers move over to the bar and yachties like Jill row ashore to join the party.

That's another word I've gotta explain. Yachties is what J&J are now that we're in British territory. The Brits don't have boats. They have yachts. And the people who live aboard yachts are thus yachties.

The BVIs are full of them. They're from all over the world. Just here in Smugglers Cove right now we have one other American couple, two British couples, a Kiwi (that's a New Zealander) with an American girlfriend, and a Danish guy who just got rid of his Swedish girlfriend and is trying out a French woman who answered his ad posted at The Schooner bar in

Roadtown, a yachtie hangout. The bulletin board there often has posts looking for sailing crew. The yachties refer to the category as Crew and Screw.

When my paint job was finally finished I felt I had graduated from lovable urchin to svelte sophisticate. There was still a lot to do before I'd be really seaworthy but now J&J decided that Jill should get a job too so that they could save more money and get away sooner. As Americans in a British colony/protectorate they had to work "under the table". Jack managed by doing "odd jobs" for AC.

Jill Gets a Job

Jill has a bicycle which she rides to a nearby village for grocery shopping. On the way she passes a cottage where she often sees a white woman sitting outside facing her ocean view and apparently writing in longhand.

She learned this was Mary Stuart, a British mystery writer whose name was familiar. One day the woman smiled and waved as Jill rode by and Jill impulsively stopped and said hello. She admitted she had never read the woman's books but was impressed to meet a published author as she was an aspiring writer herself.

The author invited her to stay for a cup of tea and an hour later Jill had a job. She would come there three mornings a week to help Mary with her current project. Jill would type and copy-edit the manuscript that was due at the publisher's in just five months. The author would never meet the deadline without her help.

Fate must have sent you, dear, she told Jill. I love working here on this lovely island but I was getting very worried about ever finding a helper. I can't type and I'm a terrible speller so your newspaper background should be just what I need. I know it's the end of the week but can you start tomorrow—Friday? At 9? Is $60 a week good?

Fantastic! said Jill. The BVI uses American currency and that was more than she had made at her full-time job in St. Croix. Jill quickly revised her shopping list to include a celebratory steak for that night.

The next morning Jill wrapped her small portable typewriter in a sweatshirt, nestled it into her bike basket, and pedaled off to her new job. The typewriter had been her uncle's, one of the few mementos she had of him and one she cherished. He had taught her to type on that old Remington, and it was one of the few possessions she insisted on bringing when she moved aboard me. Jack okayed it when he discovered it weighed just a couple of pounds and didn't take up much more space than a box of cereal.

Jill arrived on time and wheeled the bike inside as Mrs. Stuart opened the gate. As Jill unveiled her typewriter, the author grinned. What a sweet little machine, she chirped. It takes me back to my childhood! It was my uncle's Jill explained. And he died in 1952 so it goes way back. How delightful that it's a bright color Mary said. I know Jill agreed. And it happens to be my favorite—turquoise!

Well come in, come in, dear. I thought we'd sit out here today with a cuppa tea and I'll just tell you about the book and what we'll be doing to get it finished. Make yourself comfortable and I'll be right back.

Jill sat in one of the cushioned wicker chairs facing the ocean view and thought how peaceful it was. It didn't seem like a likely inspiration for a murder mystery.

Let's start with my name, the author said when she returned with a tray bearing a flowered china tea pot with dainty matching cups. I think you should call me Mary because I'm sure we're going to be friends and it's much easier. And I know you call these cookies but we'll call them biscuits so you get used to translating into British. I'm not a great cook but I do like to bake. Now, are you a mystery reader?

Not really Jill admitted.

Good said Mary. That will make you a more astute reader. Not that I expect you to edit the book, but I mean you'll be reading less for the story and more for any mistakes. By the way, I do not write blood and gore. I write Cozies. Are you familiar with the term? They're the type of mysteries that proper British ladies like me write. There's very little blood and absolutely no gore in my books. Jill said Great. I'm very glad to hear that. I have to admit I'm not a fan of blood and gore. Good the older woman smiled. You have good taste, not like a lot of the youngsters these days.

Mary proceeded to give her new assistant a short description of the story she was working on—*The Case of the Drowned Yachtie*—set in the Caribbean, involving yachts cruising between islands, and of course a mysterious death. She wasn't sure yet how that story would play out.

The main character in all of her Cozies, she told her new assistant, was Samantha Smart, a travel writer for the British travel magazine *Happy Holidays* who somehow runs into a mysterious death wherever she goes on assignment and, thanks to her contacts and her brain, was always able to solve the crime.

I love it already Jill offered. And unlike most Americans I know what a yachtie is. Actually, I'm now one of them!

What? You're a sailor? What a blessing you will be! I know just a bit about sailing so I've lucked into a very valuable asset in you!

Well, I do live on a boat now but I'm really new to sailing. My boyfriend is the real yachtie. He's American too but we have quite a few Brits in Smugglers Cove so we're learning a lot of British expressions. And of course I'm also learning another new language that might be helpful: Nauticalese.

Mary loved the word and bubbled Then you'll be doubly helpful Jill. This is such serendipity! I was a bit worried about hiring an American. You understand there could be a big problem because of our vocabularies. I write very British books of course and my characters speak like Brits so I was afraid you might try correcting my proper English expressions into improper Americanese. She smiled when Jill looked offended. I'm just warning you you'll have to study up a bit on Brit spelling and vocabulary. But your immersion into a British island is very helpful. And don't worry. I have all 12 volumes of the Oxford English Dictionary here so we can't go astray.

When they finished their tea Mary took Jill into the house. A charming little West Indian style cottage with lots of Gingerbread decoration, it was furnished simply with mostly wicker furniture and soft pastel cushions. Doors and windows were open to the tropical breeze. One corner of the small living room had been turned into a workspace for Jill with a small desk and chair.

You can put your typewriter here, dear. And there's a ream of paper in this drawer. And here are the first twenty chapters she announced with a flourish as a box full of lined-paper notebooks in various sizes and shapes landed beside the typewriter.

Don't be alarmed, dear. I'm pretty good about marking. See? This notebook is Chapter 3A. That means there will be a 3B in here somewhere. Maybe even a 3C. Here's Chapter 7. It doesn't have a letter so that means it's complete. Now don't panic when you open it up. I know I'm pretty messy but I think I'm quite legible and I've tried to make my inserts and deletes as neat as possible.

Jill was shocked to find tiny scribbles of cursive handwriting adorned not only with numerous inserts and deletes but a bevy of arrows too. Words, phrases, sentences and even paragraphs were circled and attached to arrows that seemed to go in all directions, some onto other pages, some into other notebooks!

Jill swallowed an expletive and forced herself to concentrate on a page and soon realized the puzzle was easy enough to solve.

I think I can handle it, she announced, grinning

as Mary looked very relieved.

Thank goodness, Jill. I'm sure you'll catch on quickly and I'm quite sure I'll like working with you. I'll always be close by. I write outside when it's not too hot or windy, or I'm right here in that comfy armchair by the window and I don't mind being interrupted. Well, I think you'll get to know when it's okay to intrude. If I'm gazing out the window I'm probably not in the middle of creating a fine phrase.

There was a half hour left to the morning so Mary asked Would you like to spend a few minutes looking at what you're in for? Here's 1A. This first draft doesn't have to be typed perfectly because I'm sure we'll be making changes. It's a rough draft, one that we can edit as we go along, so go ahead and cross out if you make a mistake. If you see a misspelling or grammatical error correct it but underline it so that I can see what you've changed. Here's another box for your typed pages.

And oh, here's a copy of the first book in this series. It will give you an idea of my stories and my writing style. Bye until Monday, dear. Have a good weekend.

By the end of the next week they were well into Chapter 2 and confident that Jill could do the job. Mary spent most mornings outside, writing in the shade of a large tamarind tree. As Jill typed she made notes of her questions, which they discussed during a mid-morning break when they met for a cuppa tea under the tree.

They felt comfortable with each other now and Mary revealed her modus operandi. You might not be able to detect it yet, dear, but I write by the seat of my pants. I have only a vague outline of a story when I start a new book because I know from experience that it might take off in quite different directions from my original idea. I'm not much of a planner, which works out well, because when I start a sentence it sometimes ends far away from where I originally thought it was going!

As Jill looked startled she asked Are you shocked? I know it sounds strange but I know I'm not the only fiction writer to let the story tell itself. I've read many author interviews in which the writer states that her characters write the book, that she lets her plot unfold as she goes along. The characters are usually defined by the author but they are allowed to make their own decisions. If they decide when and where they want to go and what they want to do and I think it's a delightful idea? Why then it stays. If they don't make good suggestions, then I have to do all the work!

Jill gave a delighted guffaw at that revelation. Mary grinned.

A Single Woman

Back at the cove that night the yachties were all a-twit. A 50-foot ketch is en route from Puerto Rico, planning to drop a permanent mooring in the cove. They are going to leave a friend on-board to look after the boat and they will fly over weekends on their private plane to go sailing. The friend is a young woman!

The owners love sailing in the BVIs but hate the long sail to and from San Juan, which leaves them little time to enjoy cruising if they have just a weekend. Since they also own a four-seater Cessna, they can now fly over Friday after work and have the whole weekend for sailing before flying back.

The owner/builder of the boat, Tim Sterling, had contacted Dick Conway, the boss at *AC*, who personally picked a perfect spot for the ketch—out of charter boat traffic, well protected by the reef but not too far from shore-- and dropped a brand new mooring. We were going to be the new boat's closest neighbor.

Saturday morning *Brigadoon* made an impressive arrival, clearing the narrow channel neatly and picking up the mooring under sail. As soon as the boat was secure, a big guy dinghied ashore to meet Dick and the rest of the *AC* gang. He invited them and everyone else in the cove aboard for cocktails at 5. I heard a lot of noise—talking and laughing and music—and saw a crowd of people in the big cockpit and spilling over onto the cabin top and side decks.

When J&J finally got home that night, high and giggly, they were full of exclamations about the boat—a beauty; the people—Tim and Jane and two teenage daughters and another couple, all terrific and friendly; the food—when they ran out of booze and the covies didn't show any sign of leaving, the crew threw together a bunch of canned goods and created a delicious dinner. Then the covies finally went home. Yachties are really rude, Jill giggled, we don't go home until you feed us.

But the big news of the evening was that there was a single woman named Annie who was going to stay aboard to take care of the boat. J&J were dubious. She was about their age—in her 30s, not too bad looking but quiet and shy AND she didn't know anything about boats! Uh-oh! What were they thinking!

Annie hadn't said much all evening but J&J did learn that she had recently quit her job in San Juan and didn't know what she wanted to do next so was taking a break to think about it. The Sterlings saw an opportunity to get their boat to the BVI with a live-aboard caretaker, asked their friend if she'd like to live on their boat in Tortola for a while, and she committed to at least two months. Annie had sailed with them a few times but her only skill so far, she said, was making the Bloody Marys.

Tim spent the weekend prepping Annie for her new job. This is the engine, which you have to run for an hour every day to charge the batteries to run the lights and refrigerator. This is the water tank which you have to be careful not to let run dry by collecting rain water from the awning. She didn't even know how to run an outboard motor, which she had to learn as their dinghy was so big she couldn't row it so she couldn't even get ashore! Well, actually Jill was just as clueless when she moved onto me. But she had Jack there to lead her all the way and Annie will be alone!

It was stupid of me to worry because of course Annie was the only unattached woman in a cove full of unattached men.

The first one to attach himself was a Cockney "bloke" as we say here in the BVIs. He's in charge of AC's repair shop, which is right next to the dingy dock, and so he happened to witness her first attempt to go ashore.

As she approached shore her steering went a bit askew and she ended up in the mangroves instead of at the dock. Roaring with laughter, Mac shouted you're a bit off course, lass, and waded in to rescue her. He proceeded to give her a good lesson in outboard motor navigation, charmed her with his Cockney accent and escorted her to the Bilge Bar where the rest of the crew were assembling for happy hour.

Gorda Sound

After a few weeks of making money ashore and working on me during the weekends, J&J decided it was time to take some time off and explore more of the BVIs. I of course was delighted to flex my rudders and feel some spray as we spent a few hours reaching up the Sir Francis Drake Channel, which separates the Caribbean Sea from the Atlantic Ocean.

This is what makes sailing so popular here. The channel is basically southwest to northeast, protected from the west by Tortola and on the eastern Atlantic side by several small islands including Peter, Cooper and Norman. These allow the prevailing easterly winds to offer a nice smooth ride—most of the time.

That position is called reaching and is my favorite way to sail. With the wind on my side, it propels my sails along smoothly. Steering is easy, just a little this way or that to keep the sails full, no major interruptions with flapping sails and thrashing sheets.

Sheets, you need to know, are the ropes attached to sails. Every time we tack (change the sail's directions from one side of the wind to the other) the sheet has to be loosened on one side and tightened on the other side by winding it on a winch.

Jill has read up on our destination, Gorda Sound. It's a large body of calm water surrounded by the island of Virgin Gorda and a sprinkling of protective islets.

The charter yachts naturally love this place and when we arrived we saw them anchored all over but Jack quickly found an uninhabited island with a secluded and shallow inlet where we could anchor and know we'd have it to ourselves. The charters and the other private yachts are almost all monohulls with deep keels.

As soon as we were secured and the awning was up, off came the clothes and J&J splashed into the bay. They played around for a while, then Jack grabbed his spear gun and headed out while Jill swam laps around me. We saw a local fisherman stop and talk with Jack, who soon came back with supper, an unattractive floppy fish. He proudly presented it to Jill saying Here I just found out this is named after you: Ole Wife.

That led to a playful tussle, which led to a loving tumble. They retired to the bunk below, just in case a curious cruiser came by.

Over the weekend we sailed all over the sound.

J&J visited Joe who ran a dive shop on top of Saba Rock, a character renowned throughout the BVIs for the many wives he had found and lost over the years. They went ashore for lunch at The Bitter End, the only little resort/restaurant/shop on the northern tip of the sound. And they checked out Spanish Town, a tiny village with a small marina and a few shops, guest houses and pubs.

The highlight of the Sound for J&J was The Baths. This is a stupendous assemblage of huge boulders on a stretch of beach. Jill had read they are up to 40 feet tall. They are smooth and shapely and are gracefully draped all over each other creating arches and tunnels and grottoes and tidal pools. I was wowed from my view at anchor. J&J played hide and seek as they splashed around in there. Jill said I feel like Alice in Wonderland. Jack said these are like giant sculptures.

Which led him to realize he hadn't sculpted anything but my parts for the past three years! He realized he was so focused on me it hadn't occurred to him to do anything else. But he promised himself and Jill that he would create something artistic sometime soon.

A Dog

One day when Jill was riding her bike to work she passed a dirty little dog lying by the side of the road. He lifted his head hopefully as she said Hi Pup while pedaling past. He was still there three hours later when she was riding back home. She stopped and approached him warily. The dog rolled over and whimpered. One leg was bloody and limp. Jill looked around. She was on an open stretch of road, no houses or people nearby. The dog was very thin and unkempt and did not have a collar nor any other sign of domestication. She decided he was a stray, not uncommon on the island, and decided to take him with her. Jack, who seemed capable of fixing anything, could maybe fix this mutt.

Jack was down in the bilge of one of the smaller charter boats, a 32. He was happy to come up for air when she called and didn't seem unduly surprised when he saw Jill holding a very sorry-looking mutt. He gave her a hug and gently took the dog into the workshop.

He felt the damaged leg, said OK Buddy when the dog yelped, then It doesn't feel broken, just cut and bruised. He sent Jill into the office for first aid supplies, then dressed the wound. He poured some water into a nearby bucket and held the dog while he drank—and drank. He said I guess we'd better keep him till he's healed.

Jill nodded, her eyes brimming, thinking what a guy, he was so kind and gentle—and capable. The dog looked grateful too. They used Jill's bike basket as a carrier, lining it with kapok from the boat cushion corner. The mutt was settled in the basket, yelping, and Jill carried the bundle to her dingy. She steadied it between her feet as she rowed home.

I had never seen an animal before. This pathetic mess was my first introduction. I was kind of disgusted, but fascinated. First Jill fed him some small bites of bread, which he gobbled. Then he got a bowl of water that he slurped rudely. Then she put some of tonight's hamburger meat into a bowl and looked at him proudly as he practically inhaled it all. I guess you're feeling a little better now little one she cooed. Yuck! What's the matter with Jill! She's giving away our best food!

Well, as you can probably guess, the dog hasn't left yet. In a couple of days he was able to limp on all fours. Soon he was able to climb the companionways so now he's all over the place, in the galley when Jill is cooking of course, and on their bunk when they go to bed! One or both of them takes him ashore at least twice a day. Well I'm okay with that. I personally found it disgusting when he was peeing and pooing on my decks!

He's about a foot tall and maybe a foot and a half long. They figure he's maybe a year old and they hope he's finished growing. He's got lots of hair and it's all colors--black and brown and white and yellow. They figure he's a mix of every kind of dog on the planet. He keeps busy during the day watching all the boats come and go in the cove. If a sailboat or a dingy comes close to us, he goes berserk! He barks! He says WOOF! Woof Woof WOOF! And can you believe it? Jack and Jill call him—affectionately!—Woof. Or Woofer. Or Woofie Doofie!

Sure I'm jealous. But I'm working on that. They still treat me very well and I still feel a close bond with both of them. I get plenty of attention. And affection. Just not quite as much maybe as before the Woofer. I'm afraid I might have the new-baby-in-the-family syndrome.

Captain Jack

Jack is so-so about his job. He's fine with the work. He likes the variety. One day he's overhauling an engine and the next fixing anything from loose screws and scratched varnish to a leaky head. But regular hours? Every day?

So when there was a last minute need for a charter captain, he jumped at the opportunity. Two middle-aged couples from Boston had chartered a boat for a week and the freelance captain who had been hired for the job was suddenly unavailable. It was high season and all the other captains were fully booked.

Dick, the Antilles boss, in desperation asked Jack. He figured that even though his scruffy handyman didn't look much like a captain and wasn't trained for the job, he certainly knew the charter boat inside and out and he knew the islands as well as anyone—and, he was good company. Jack of course was delighted for a chance to sail for a week and get paid for it! He had a day to get ready.

Jill was happy to see him excited. She cut his hair and beard as neatly as she could, scrubbed his Antilles Charter T shirts, and dashed into Roadtown to buy him a decent pair of shorts. He didn't own any that weren't torn or worn to shreds or covered in paint or glue. She decided he looked presentable enough to be a charter captain when she kissed him goodbye.

It was her first time alone on me overnight and I worried she might be lonely. She did talk to me more than usual, which I loved. She told me how much she loved me and Jack and how happy she was to be part of our lives. She puttered around like a little housewife on her day off, wiping my bulwarks and swabbing my decks and rearranging the cushions on my deck. She usually spends her days aboard at her typewriter or cleaning up whatever mess Jack left the day before. He always manages to tinker with something when he comes home after work.

Two days later, while Jill was typing in the salon, we heard an outboard motor approaching. She popped her head out of the hatch and saw Jack! Oh lord she wailed, something awful happened to the charter. But no! Jack leapt onboard and said Get your gear together Tweetie. You're going sailing too!

(Have I mentioned that Jack calls her Tweetie now? It's because she squeaks a bit when she's excited.)

It seems the very nice people from Boston had told him they felt bad leaving him alone on the boat when they went ashore for dinner in the evenings and asked if he was married. He said no but he had a girlfriend. They said why don't you bring her along? He said she's got a job and she's looking after our dog. They said see if she can get a few days off and bring the dog! You're making this up said Jill.

No. Look over there. There was the charter boat and there were four people on deck waving at them and beckoning them over. Wow said Jill. This is unbelievable! Okay! Did you hear that, Woof? We're going sailing on an AC 44!

They scrambled to pack a bag with some dogfood and clothes for Jill, then quickly closed all my portholes and hatches. Jack patted me on the port cabin top and said Be a good girl *Hula* and they were gone.

Woe! I had never been left alone before for more than a few hours. This would be a couple of days! What if there was a storm! What if the anchor dragged! What would I do with myself?

I soon pulled myself together and decided to enjoy the peace and quiet. This was a lovely cove, I had views of the reef, the channel and a distant island one way, the hillside with three pretty houses another, and the busy *AC* docks ashore. And I knew that the *AC* guys would take care of me if there was any trouble. They're a great bunch of guys. I relaxed and enjoyed the restful silence, the swish of tiny waves along my sides. I'd never really thought much about my relationship with the water, except when we were sailing and I got lots of slapping and splashing. This was nice, just me—a turquoise catamaran— sitting in a gently lapping turquoise bath. Aahh!

The Hippies

Our peaceful little anchorage is getting quite a

shake-up. Something looking spookily like a ghost ship has dropped anchor in the cove and deployed a cast of characters straight out of *Hair*. A nice old three-masted schooner that has seen better days, *Aquarius* has a young crew of both sexes who are very casual about covering up. They're anchored close enough to us that we are often treated to the sight of bare skin. I think there are seven aboard.

The first few days they stuck to themselves, just hanging out, smoking and playing music, mostly from *Hair*. Jack and Jill both love *"Let the Sun Shine In"*. When they hear that wafting over the water they both chime in. We are downwind of *Aquarius* so also get the occasional whiff of marijuana—and enjoy that too.

My people don't smoke themselves. Jack never has and probably never will. He doesn't drink a lot either. He's happy with a beer or two every evening. At parties he might drink a bit more. Jill smoked pot at parties before meeting Jack, but since being with him she doesn't. Now she just drinks wine—and is really fun when she's had a few. We rarely see a joint in Smugglers Cove. Sailors are traditionally drinkers, and the 60s apparently didn't change that.

Well, the captain finally came over to introduce himself to us, his nearest neighbors. He's from the West Coast but found the old schooner abandoned on the Gulf Coast when he was there for a rock concert. He got a few of his buddies interested in helping him salvage it with the lure of sailing to the West Indies. He turns out to be a nice guy. J&J liked him. Jack said he and his crew should all go ashore for happy hour. Thanks but none of us are drinkers said Blondie.

Yes! That's what he said his name was. It figures. He does have a head of beautiful long yellow hair. We can't imagine anyone sailing with that blowing around his face so were relieved when he pulled a rubber band out of his pocket and tied it back. We've been supporting the rubber band business since we went to sea, he announced. We all have hair we don't want to cut.

Jill suggested that if they didn't want to come to the Bilge Bar maybe they'd like to join the group Friday night when all the covies go into Roadtown for pizza. We have the use of the AC Land Rovers she said. Far out said Blondie. We'll probably do that.

He asked for a look around, said it looked like an unusual boat, so Jack gave him the tour and had to explain a lot. The hippie doesn't know much about boats and asked a lot of questions.

Late Friday afternoon a few of the hippies squeezed into a truck bed with Jack and Jill and some others and introduced themselves by passing around a joint. Cockney Mac suggested they sing, and *Let the*

Sun Shine In accompanied them up and down the hills and around the curves into town.

Pizza night at Gloria's Glorious Grill was really jumping that night. When a hippie stepped outside for a smoke he was often followed by a local or two and pretty soon everyone was happy high and singing songs from *Hair*.

The hippies didn't stay in the cove much longer. They were island hopping and had a few more to see before they ran out of money.

The Cozie

Jill is having a ball typing up *The Case of the Drowned Yachtie*, which is the third in Mary's current series of Cozies.

Yes, of course I listen in as she babbles on to Jack about it all the time. I can tell you that so far it looks like the yachtie's going to meet his fate on Norman Island. That, Jack tells us, is often thought to be the Treasure Island of Robert Lewis Stevenson's classic story.

Yes says Jill. But our victim is looking for a different kind of treasure. He's a geologist and he's studying the volcanic rocks on various Caribbean islands. He's sailed his way up through the Windwards and Leewards and is finishing up here in the Virgins. I don't know yet why he picked Norman and I don't know how or why he drowns. I haven't got to that part yet. I'll let you know. I'm not sure Mary knows yet either.

I know, Jill answers Jack's raised eyebrows. It's weird but she says that not always knowing where the plot will go next works well for her. Remember *The Case of the Strangled Bird Watcher?* We read that a couple of weeks ago? Well, Mary says he was originally going to be bonked by a coconut—that story also took place in the tropics—but the murderer changed the weapon to a more appropriate weapon. Remember? Binoculars? Strap around the neck?

A few days later Jill says Hey, Mary's got some good research on Norman Island but since she's right here across the channel she would like to go there. Why don't we take her over some weekend! Good idea says Jack.

It'll have to be a day sail says Jill. I doubt she'll want to sleep aboard. I'll warn her we don't have exactly posh accommodations.

It turns out Mary has sailed on all kinds of boats around England and in various places around the world where she has gone to find stories, so she's interested in sailing on me! Her first catamaran. She loves to snorkel and is willing to learn to dive if she decides her story requires it.

We haven't been to Norman yet so Jack and Jill and I are eager too.

Treasure Island

Mary was a delightful guest. Jack was happy that she was interested in everything about me and even asked to take the helm briefly. She loved the fact that we didn't lean over. She could actually walk around the deck without hanging on to something. It was a perfect sailing day and we had a nice brisk ride across the channel to Norman. Norman is an uninhabited island, so it's lovely and pristine.

We went to Privateer Bay and anchored as close as we felt safe to the rocks off Treasure Point. That's where there really was a treasure chest found buried in a cave ashore hundreds of years ago. Mary wanted to snorkel in the water near the cave where she thought her victim's body would be found floating by a passing yacht.

The water is deep there so we had to pay out a lot of chain but we had the place to ourselves. Any other yachts at Norman that day were anchored on the other side of the island.

Right away all three of them donned their masks, snorkels and flippers and jumped off the side deck into the water. They headed toward the high rocky point, which they found surrounded by large rocks, some totally underwater, and a strong surge. Jack signaled they shouldn't go any farther. Mary took the snorkel out of her mouth long enough to say I won't be a minute and charged ahead. Alarmed, J&J felt they had to protect the older woman so followed. She swam right up to the rock wall, gave it a good feel, then dove for much longer than a minute between several of the adjacent rocks. When she finally surfaced for good she asked Jack to carry a bag now heavy with small rock specimens, then they all swam back to me. Safe and sound. Whew!

They had a big and long leisurely lunch. Jill had made lots of sandwiches and Mary had brought a bottle of wine. She was all wound up. I think I've got the climax now, she announced. My geologist couldn't scale the rocky point so he snorkeled around the bottom, collecting smaller bits of rock he found there and putting them in a big sack. My killer is a local fisherman who has been watching him come and go from Treasure Point for a few days and deducted he must have found some leftover Pieces of Eight buried in the sand. That's what is said to have filled the treasure chest found in the days of piracy. Mary claimed she would return the rocks once she finished the book.

Brilliant said Jill.

What a way to write a book! Said Jack. You just made that up!

You're right said the author. I don't bother to plan much of a plot because for me the story unfolds as I write. I might think it will turn right at a certain point but when the time comes it's obvious it has to turn left. So there's no point in me wasting my time plotting. I know, she said to Jack's you've got to be kidding look. Luckily I have a publisher who's just as eccentric as I am. He enjoys my surprises, she giggled.

Captain Jack

It's the height of the yacht charter season. Jack has become a popular captain so is often called on to drop his handyman tools and pick up his captain's hat.

That's just a figure of speech, of course. He wouldn't be caught dead in anything pretentious like that. In fact he wears the same old floppy cotton bucket hat he always wears. Though Jill does scrub it with bleach every time he has a charter.

Dick Conway, his boss, had hoped to spruce him up in the beginning but backed off once he realized the charterers were enchanted with Jack just the way he was. Jill and I knew he was a perfect captain—he knows exactly what to do no matter what happens AND he's cheerful, funny and kind. Even when Jill does something wrong he's perfectly calm and nice about it. He just calmly tells her how to do it right. It's nice that Dick recognized he shouldn't improve Jack!

Captains make more money than maintenance men, and Jack usually comes home with a nice paycheck from AC AND a tip from the charterers. He's really enjoying that. And having nothing to do but sail for a week, eat gourmet provisions including steak dinners, and almost always be with clients who think he's great.

But having some money in the bank now is so exciting he can think about the Pacific again and his lifetime wish to circumnavigate the globe. He knows he will have to work his way around the world, but now that he's gathering a nest egg for emergencies, the reality really seems within reach.

I'm ready. I guess. I mean that's the reason I exist. But who knew I'd enjoy what we're doing right now so much. Going to sea occasionally, resting for long periods in beautiful warm anchorages, enjoying the daily camaraderie of Jack and Jill and Woof. Yes, him too. Most of all I guess my problem is Will Jill go with us? Will the Woofer?

Jill's Backstory

Jill is from a small town in Connecticut. Her parents aren't as well off as Jack's and they don't live near the coast so Jill did not grow up sailing.

She went to UConn, the state university, and majored in English Lit. She spent her junior year abroad in France, thinking she might meet up with a Hemingway/Fitzgerald type of literary crowd, but that didn't happen. She took one writing class her senior year, found she had a gift for it and decided that would be her life.

I'm sure she could have written a fabulous novel right away but of course after college she was expected to make a living. A major in literature and a desire to write didn't open any doors for her but she finally got a job on a sad little weekly newspaper near her family's home. She had hoped to be on her own after college but since the job was just a few miles from her parents' house it made sense to move back in. And of course it's always nice not to have to pay rent.

The weekly paper was run by an elderly man who had inherited it from his father. Sam had been doing it by himself for years but was running out of steam and needed some help. An eager young woman familiar with the area was perfect. Jill became assistant editor of the *Country Crier.*

She survived the school board meetings and other boring gatherings she had to cover by digging up exciting feature stories hidden in the countryside. That in turn revived the area's interest in the weekly, and both ads and subscriptions picked up noticeably.

Her first coup was discovering that the author of a series of light-hearted novels often on best seller lists was actually a reclusive middle-aged woman who had recently bought an old house in the area.

Jill stumbled on the story, she jokes, by turning her ankle while out running on a dirt road in the woods. She spied a house through the trees and hopped to the door, asking to use the phone so she could call her family for a ride home.

While waiting for her dad to pick her up, she saw a stack of books on a table and said Wow! I see you're a big fan of Patsy Pride!

Oh. Yes. The woman replied.

Jill said I have to admit I haven't read any of them. Can I look at the books while I'm waiting?

A nod OK.

That led to her first "exclusive". As Jill leafed through copies of seven novels, all signed by the author but otherwise pristine, she somehow guessed that the uncomfortable woman fidgeting beside her was likely personally involved.

It's none of my business she said but may I ask? Are you by any chance related to Miss Patsy?

No no no!

Sorry. It just struck me as odd that the books look new. Like they haven't been read at all! I was sort of hoping you had some kind of connection. I have to admit I work for the *Country Crier* and I was hoping there might be a story here.

A car horn broke the awkward silence. Oh there's my dad. Very sorry for the intrusion. Thanks for the use of the phone. Bye she said hopping to the door and down the steps to the car.

Believe it or not, Jill says the woman actually phoned her at the paper later and invited her to come back. I'll give you a story after all she said. That's the dullest paper I've ever seen and I'd like to give it some spice.

Jill looked up that house's sale at the town hall and discovered the new owner was Margaret Percy, 45. The name didn't mean a thing but she was game for any chance of a spicy story.

Mrs. Percy aka Patsy Pride was much more cordial on the second visit. Sorry I wasn't more polite the other day she said. You took me by surprise and I'm not used to people dropping by. I'm almost a recluse. But you seem like a nice girl and it would be good to give that rag you work for an interesting story. She proceeded to do so. Here it is!

Famous Author Hides Out Here

By Jill Sanford

Did you ever wonder if the popular author Patsy Pride was a pseudonym? Did it ever occur to you that the best-selling writer of the Small Town series might be a quiet woman who worships silence and lives near you? Well, believe it.

This reporter has pledged to keep secret the real identity of the author and her exact location but we can reveal that the following story has been verified by the books' publisher and permission to print the facts has been granted to this newspaper only by both the author and the publisher. Under penalty of (you name it) if the pledge is broken. Patsy Pride (we'll call her Mrs. X) says she is signaling readers not to take her stories too seriously. To "just enjoy the romp!"

Mrs. X is a divorcee who lives alone in the friendly woods of our county. Her reclusive lifestyle has succeeded in keeping her writer identity a secret.

She says she started writing when she realized that her extremely handsome and well-mannered husband was, in fact, "a devious, conniving, heartless, soul-destroying son of a bitch!"

To help her get through the discovery that he had disappeared with all the profits from their arty gift shop in New Mexico she started to write a journal. It became a diatribe about HIM, how he used his sexy charm and her business acumen to create a delightful small business which made a small fortune. When it was so good they agreed it can't get any better than this he suddenly cleaned out all their savings and disappeared, leaving her with large mortgages on the shop and their home and a huge hurt.

She found she loved writing in the journal and had a flair for it so she decided to try writing a book. She said "Having dumped all my hatred and hurt into the journal I just wanted to switch gears and move onto something funny and fun."

Enter the Small Town series and its "town crier", Billy Sunshine. "Billy is the exact opposite of my bastard husband," she explained. "He's a simple sweetheart who lights up the town with his funny jokes and kind deeds."

Mrs. X has so far written seven of these popular stories, which are lauded as "lovable" and "delightful" by most critics and have often appeared somewhere near the middle of many Best Seller fiction lists. "I'm delighted to be there," she admitted, and I'm really happy to know there are so many people in the world who still enjoy reading about good people. It's incredible to share The List with the flashy writers like Norman Mailer and John Le Carre."

Mrs. X says she loves our county and the privacy she has

found here. She wants to continue being "unknown."

Signed copies of her books are available at Reader Feeder on Main Street, Centerville. Mrs. X said she has given that special perk to the local bookstore as a gift to her "lovely new neighbors."

#

Wow. She read that to Jack the other night and he was so cute. Oh Tweet he said. You're really good. I mean I knew that, you wrote that super piece about me and *Hula* that I thought was terrific but I never stopped to think that you could turn out something that interesting all the time. And here I am making you paint and clean when you could be writing best sellers too!

Oh no Sweetie she said. Yeah, after he started calling her Tweetie she now calls him Sweetie. Because, she says and I agree, he is. Anyway, she explained that she loved working on me, that this was her home and she loved sharing me with him. Yes she loves to write too and is so glad that he thinks she's so good but right now she wants to enjoy us fully. She'll write when she has nothing better to do.

What a gal. She's so smart and lovable.

None of the yachties around here have any idea what she can do. To them she's just Jack's girlfriend.

Wait til she finishes that novel she's writing so quietly. I know it'll be a best seller. I hear her yahooing and guffawing when she's produced a specially good line! I like that it might be funny!

First Cruise

We're finally going cruising!

Jack announced it last night. "The season" is over-- the winter months when the statesiders come down to enjoy a tropical sailing vacation is over. Jack feels rich, thanks to all the nice tips from the charters he captained, and he's raring to go beyond the Virgin Islands, to really go to sea! That's why he built me, after all, so I'm psyched too!

Actually I'm a little nervous. I mean I know that Jack did a beautiful job creating me and I certainly feel strong and capable but, you know, I've never even been sailing at night, I've never been underway for more than a day at a time. Now we're talking about island hopping around the Caribbean. That will include overnight passages! Two day passages! Maybe even more!

I know. I know. Eventually we're going to cross oceans. Never mind mere overnights. It will be days at a time, maybe even weeks. We certainly need to practice.

Jill is up and down. Oh my god *Hula* she said after Jack broke the news. Hear that? We're going to see the rest of the Caribbean! Jamaica! Trinidad, St Barts! All these places I've wanted to go to!

But then a little later, after Jack went ashore to fix a leaky bilge in a 44, she said Oh *Hula*, what are we in for? We'll be out there in the open sea. We'll be in strange harbors. We'll be in foreign countries!

Woofer's all excited because they are! Uh-oh. I wonder what they'll do about him.

Jack told Dick the boss that we would be leaving in a few weeks, that he would finish up any necessary jobs first but also needed time to spend on me. Dick said Jack was the best employee he ever had and there would always be a job for him if ever he wanted to come back.

Jill told Mary she was leaving but the timing is perfect. *The Drowned Yachtie* is almost finished so Jill will have just enough time to finish typing it before we go. Those two are a great pair now. They know they will meet again. Jill has promised to keep Mary informed of our itineraries and Mary is sure we'll be cruising in other places where her amateur detective Samantha will love to solve a crime! She's already planning a story in Australia, knowing that's high on Jack's must-go-to list!

Jack is taking advantage of the AC facilities to get me ready for sea.

It's hard to believe but he's put together a 50 horsepower outboard motor by attaching a Mercury top salvaged from the bottom of the sea to a Johnson bottom found in a junkyard and a three-foot stainless steel shaft he fashioned himself—without benefit of machine tools. And it works! We had a spin around the cove the other day! What a genius!

His pals call him Rube Goldberg. In case you're not familiar with that name, he's a legendary nutcase who makes wildly complicated contraptions out of the most unlikely materials—things that miraculously work! Jack's gotta be a direct descendent of his!

This will surely make life easier when we have to motor. Jill is ecstatic. She gets nervous when we have to maneuver through anchorages. With our old 7 horse Chrysler we definitely didn't turn on a dime.

He's also using the AC's sail sewing machine to remake a sail—an old genoa that was left by a cruiser who had a new one made in Roadtown. We wanna make sure you have a complete wardrobe when you go out into the world *Hula* he told me, explaining that a genoa is a big sail that replaces the jib when the wind is light. Or is rigged side by side with the jib when the wind is coming from directly behind and we can sail "wing and wing". That sounds terrific. I know I'll love sailing downwind!

The rest of my wardrobe—that's what they call the sails, really—was also re-made from hand-me-downs but that was all on Jack's old machine, the one he bought from a brassiere factory in Puerto Rico. As you know we don't have electricity so J&J crank it by hand.

He's also making new hatch covers that can be locked. We've never felt the need before but Jill pointed out we wouldn't always be in idyllic little harbors like the Virgin Islands. So now we have locks and keys.

***

The last few weeks in Smugglers Cove were bittersweet. It was the first place we had all lived together and after more than a year it felt like home and we were sad to leave but we were embarking on a circumnavigation of the world! That's a huge thought and certainly something to be excited about.

Don't think of it like that Jack advised Jill when she started blubbering about the wild blue yonder and could she handle it and should she go so far away from her family etcetera. You'll have plenty of time to see how you handle it and we're not going that far yet. We'll probably be at least another year in the

Caribbean. You can easily fly home whenever you want to.

The plan is to circle the Caribbean Sea. Jack now thinks we'll probably stop at most of the islands. There are so many! Jill has done some homework and here's the complete list. Since we're near the northeast corner of the sea, we'll start by heading south, down island as we say here in the Virgins.

THE LESSER ANTILLES

Leeward Islands

(Anguilla, St. Martin, St. Barts, Saba, Statia, St. Kitts, Nevis, Antigua, Barbuda, Guadeloupe).

Windward Islands

(The Saints, Dominica, Martinique, St. Lucia, Barbados, St. Vincent, the Grenadines, Grenada).

Can you believe all those islands? That's just the beginning of our cruise and it sounds like Jack or Jill or both of them want to see each and every one of them!

Read on.

NORTH COAST OF SOUTH AMERICA

Trinidad and Tobago

The Venezuelan islands: Margarita, Tortuga and a few tiny un-named specks

The Netherlands Antilles (the ABC islands of Aruba, Bonaire and Curacao)

We'll skip Columbian waters because there's a lot of trouble going on there and we're skipping the western side of the Caribbean for now. That's Central America, which we'll probably explore later—on our way to the Panama Canal!

Instead we will scoot directly north to

THE GREATER ANTILLES

The Cayman Islands

Jamaica

Hispaniola (Haiti and the Dominican Republic)

Puerto Rico

Jill, whose Spanish is pretty good, is sorry we can't include Cuba. Castro doesn't want us.

Whew. That sounds like a lifetime of cruising to me, but according to Jack we're just getting started. By the time we've done the Caribbean thoroughly we'll be seasoned cruisers and ready for whatever the rest of the world has to offer.

I like that most of these will be short passages and that it should take us a long time to cover all that territory. Jill and I aren't ready for big oceans yet.

Uh-oh. What about the Woofer? Yes, he's going with us. To start anyway. If it doesn't work out for him we'll find him another good home along the way Jack said. Oh Woofer Jill said please be a good cruiser.

We've gotta stay together.

The Anegada Passage

Oh my god, what a way to start. The Anegada Passage. Jack apparently got a briefing before we left that this would probably be a wild crossing and thank god Jack's buddy Bruce decided he wanted to go to St. Martin with us so Jack had help. Jill and the Woof were certainly useless! This was one hell of a trip!

This is a very deep trench and we were heading slightly southeast, directly into the wind, and the wind was very strong. It was very rough all the way and we were out there for 36 hours, including a long dark and scary night trying to tack this way and that but being tossed around like crazy!

I'm proud to say I handled it well. My hulls had never been whacked so hard and those waves never let up but I stayed strong. Jack sure knew what he was doing when he put me together. Those rubber mounts attaching the crossbeams got quite a workout but they kept me upright and let me keep plowing through wave after wave. I admit there were moments when my bows took a dive that seemed endless but I was always able to surface without too much panic.

Now I understand the term Shakedown Cruise. I

know it's meant to refer to a first time out when new owners are testing all the new equipment. But for us it was literal. We had an honest-to-god shakedown on our first real passage.

First we heard a lot of crashing below. Jill was hiding out in the port cabin, where the galley is. Jack opened the hatch a crack to make sure she was okay. She was gathering up all the stuff that had bounced off the shelves. I'll take care of this she said. She stuffed everything loose into the built-in lockers.

Meanwhile all hell started breaking loose on deck. The boys quickly lowered the mizzen sail and reefed the main. That means they shortened it, dropping it maybe halfway and tying it down to the boom.

Ohmygod! What a scare. They could have been knocked overboard by the boom or killed by that sheet snapping back and forth. Thank god they knew what to do, that they kept their wits about them and did what had to be done. They were like a couple of old salts. When the crisis was over they did take a minute to shake hands, but otherwise acted as if nothing terribly special had happened.

I am so impressed with my captain. He's never been in a stormy sea before but he seems to know instinctively what to do in every crisis. A born sailor.

We slammed around for a few more hours,

making little headway. The storm jib was the only sail still up but by midnight both guys were on the bow lowering even that. Another life-threatening feat! Then we were 'hove to' for a few hours, not fighting the sea any more, just letting the waves toss us around. That was not a great feeling but it seemed almost peaceful by comparison. I found my own rhythm, riding the waves, not fighting them. The wind seemed to drop several knots.

Dawn was never more welcome. It seemed that the lighter the sky got the tamer the sea became. Now it was just plain rough, not stormy. The reefed main went back up and eventually the other sails.

And finally Jill actually emerged from the port cabin! All night she did no more than poke her head out to hand sandwiches to the guys and have a horrified look around, then quickly drop below again and shut the hatch. She did that a few times, and the boys were so grateful!

We found out later that Jill didn't have foul weather gear. She didn't even know what it was! And it never occurred to Jack that she didn't have any so there she was in a stormy sea totally unprepared. It's a good thing she wasn't needed on deck and could hide out below.

I'm so proud of her! She told me later that she was scared to death, especially when she saw the guys wrestling down the jib with the waves crashing into them, and she thought they might not survive.

This was the worst possible way to start her cruising life but she didn't freak out. She knew she couldn't help on deck but she could make sandwiches and she did that. Feeding the guys was an important contribution and they appreciated that. They may not have made it without her.

When she finally emerged—with hot cocoa!—she got a rousing cheer from the boys. And me.

A few hours later we dropped anchor in Simpson Harbor, Saint Martin. We were bona-fide cruisers now!

Woofer flew up the companionway when we entered calm water and he sensed we were nearing land. Jill said he'd been cowering under the salon table all night, shaking. When he heard the anchor chain being readied he ran to the port bow and sniffed ecstatically.

Once we dropped anchor Jill immediately launched her dinghy to take him ashore. Jack said wait, wait! You can't go ashore til we clear Customs! I know said Jill but poor Woofer, he's been holding everything in for 36 hours! I won't be a minute. (That's the promise Author Mary made before diving for about 20 minutes and J&J have been fooling around with the phrase ever since.)

The good news is that the Woof did relieve himself and they were back on the boat in no time and no one reported us to the authorities. Jill made another batch of peanut butter sandwiches, a rum bottle made the rounds several times, and everyone collapsed for a few hours.

St. Martin

Bruce couldn't get off me fast enough. He is visiting friends on the French side of the island for a week, then flying back to Tortola. He vowed that wherever he went next it would not be across the Anegada Passage.

We're in Simpson Bay on the Dutch side of the island because there's a fairly famous catamaran builder whose shop is just around the corner in Simpson Lagoon. Jack is dying to meet him and see his boats.

I'm upset. Those boats will be custom-designed and built by a genuine shipwright. They will be beautiful and have all the bells and whistles. They will laugh at me—an ugly duckling homemade from do-it yourself plans with no bells or whistles at all.

Jack's so kind I know he just doesn't realize that I have feelings too. I'll pray he doesn't fall in love with one of those beauties and ditch me. Oh, I know he can't begin to afford anything like them so I'll just pray he doesn't lose respect, or affection, for me.

Now we're getting ready for company. Jill's best friend Cookie is coming with her boyfriend Mike for a week or so. Jill hasn't seen them since she left St. Croix over a year ago. Jack hardly knows them. We'll do a little cruising with them.

We've gotta get you spruced up *Hula* Jack said. You look like you've been through the wringer. That's a pretty good image Jill said. I'll add that you went through the washer too. Anegada was a full service laundry. But we'll get you nicely rinsed and dried now.

They spent a couple of days cleaning me up and getting me ready for our first overnight guests. J&J usually sleep on deck unless it's rainy or too windy so Cookie and Mike will use their bunk below. That's really small—not much bigger than a single bed Jill says—so they're clearing off the forward bunk too. Mike is really tall and Cookie likes to spread out so they might want separate bunks.

Jill's pleased, because that bunk's always been covered with what she calls Jack's junk and he calls essential parts. She figured it could all go into the storage space under the bunk but guess what? That's already full of other essential stuff.

There's a very small dusty V-shaped cubbyhole

at the end of the bunk and that's now holding the loose essentials. Both bunks are now equipped with bedding Jill found at the thrift shop and washed twice at the laundromat. Our guests can have the starboard cabin to themselves.

I don't think I've mentioned that Jack has worked some magic in the salon, which is actually just a table and bench under the port deck. He's made the table height adjustable so that it can be dropped down to bench height to make another bunk! Just in time. Cookie and Mike can have J&J's bunk and in case they don't sleep on deck J&J will sleep in the salon.

Our salon's main purpose is actually navigation. That's where Jack keeps his charts and sextant and that's where he'll work out our position when we start to use the sextant. So far we've rarely been out of sight of land.

His sextant is predictably unique. It's actually designed for aviators but since he was given it for free and marine sextants cost a bundle he's decided why not—the celestial information they gather is the same.

Cookie and Mike wanted to go cruising of course but they had just a few days so we went only to the closest island, Saba, planning to spend a few days there. What a place. It's just a huge volcanic rock jutting out of the water. But settlers managed to work a sort of village into the hillsides and there's a small population there. There's even a tiny air strip dug into the side! But there's absolutely no harbor and the water all around is deep-deep. We circled the island and found a spot where a few fishing boats were tied up to a pier so Jack tied us up to one of them. The crew managed to get a guided tour of the island which is so cute Jill said I gotta write a story about this. We spent the night banging hulls with the fishing boat and took off for St. Martin early the next day.

It's just a short sail back to St. Martin but it was kinda eerie. The sky was a spooky yellow-gray and the wind was on-off. We were glad to get back to Simpson Bay. It's a small anchorage and not popular because there are a couple of small reefs so you have to anchor carefully between them. But Jack likes it because it's convenient to the lagoon. He's already walked across the bridge a couple of times to meet Hans, the boat builder, and we plan to move in there for a while when C&M leave.

Ohmygod what a return! That eerie crossing was the preamble to a tropical depression which later became a tropical storm then a hurricane by the time it reached Haiti. My crew had gone ashore for dinner at the Oasis, a nice restaurant that's the only building in Simpson Bay.

By the time they got back to me the bay was very rough and it was very windy. They still didn't know what was coming—no one had listened to a weather report—but Jack wisely decided to put out another anchor and to hoist the dingy aboard and raise the boarding ladder. They cleared the decks of anything that could fly away and tied down anything that could be ripped off and retired below, C&M to the starboard cabin and J&J to the port, each with a bottle of rum.

It was another wild night. A bit like the Anegada Passage except instead of being tossed around by waves we were jerked around by our anchors. Jack was so worried about them holding he spent most of the night on the companionway, hanging on tight to the hatch cover as he opened it a crack to look around. He couldn't see much but swirling water in the air.

Jill was able to stretch out some on the cushioned salon bench and she kept trying to get something on the radio but there was nothing there but very loud static. She managed to doze off a bit near dawn, when the wind subsided a bit and she was no longer bouncing off the cushion every minute. But my captain was totally alert every second and I felt confident he would keep us from harm.

He finally relaxed a bit when it was light enough to look around and he could see we hadn't dragged anchor at all and nothing was in danger of hitting us. I guess Jill had been holding her breath. As soon as she heard the good news she hugged him hard and burst into tears.

Ohmygod she said. Thank god you knew what to do. Then ohmygod we've gotta check on Cookie and Mike! They're probably a mess!

Let's send them an invitation for coffee Jack said. That guy, he always knows how to make her lighten up! Jill picked up pencil and paper and wrote a formal invitation:

Mr. and Mrs Port Cabin
Request the presence
Of Mr and Mrs Starboard Cabin
For coffee and crumpets
Now
Dress Optional

Jack took the paper and opened the hatch cautiously. It was still very gusty and we were still lurching a bit but it had almost stopped raining. He crawled across the deck, opened the starboard hatch a crack and yelled Mailman!, dropped the note and closed the hatch and dashed back.

By the time Jill had made coffee and unwrapped some local buns Mr. and Mrs. Starboard had staggered across the deck and scrambled down the port ladder. Cookie was attired in Jill's favorite long slinky dress and Mike was wearing Jack's only dress-up garment, a sports coat.

What a crew! They all know how to have fun, even

after a life-threatening night!

It took all day for the storm to pass completely and everyone spent the day recovering and cleaning up the mess. Once again everything on the shelves had been tossed off despite the fiddles (strips of wood) Jack had so carefully attached across them. He swam ashore with Woofer, who hadn't peed or pooed in at least 24 hours.

We still couldn't get anything on the radio so next day Mike dinghied ashore to call the airport. The storm had probably hit St. Croix too and he wanted to get home to check it out.

The phone at The Oasis was down so he decided to hitchhike to the airport. Since he was wearing only his Speedo under his foul-weather jacket he looked naked so most cars passed him by. He finally got a ride with, he claims, a beautiful Dutch nurse. No planes were going in or out that day but he got tickets for a flight the next day. He heard that two yachts had washed up on the shore in Philipsburg Harbor. I guess it wasn't so stupid for us to anchor between two reefs after all!

Simpson Lagoon

This is a Sleepy Lagoon. The first night we were in here J&J were goofing around singing that song. Jack tried to croon like Bing Crosby so Jill aped Doris Day. They limped through the lyrics with a bunch of tra-la-las and made it to *A tropical moon, a sleepy lagoon and youuu* followed by shrieks of laughter and hugging and kissing.

I was thrilled. Jill's been a little mopey lately. Well, with good reason. I mean the Anegada Passage? Then a tropical depression? She had no idea how bad it could get and she had to find that out right away? Who wouldn't be upset! She cheered up having Cookie here and now I hope the sleepy lagoon will restore her.

It really is sleepy. It's smooth like a lake and a nice pale blue. There are a few other boats anchored here but I think only one is occupied—by one of the workers at the boatyard and his family. The only buildings ashore are like Quonset huts. That's where they make the multihulls—not just catamarans but trimarans too.

Jack's been ashore a lot helping Hans put the finishing touches on the catamaran that's being launched next weekend. I'm happy to say he's not enamored of all the bells and whistles. The owners will live aboard and will have all the comforts of home, including a freezer! They're going to be using it as a day charter so it's got gear Jack wouldn't think of buying, like a windlass for the anchor chain and self-winding winches.

Jill has no urgent boat chores and she doesn't feel like writing. The novel isn't going anywhere so that doesn't help her disposition. She and Woof have taken the same walk around the lagoon many times. She doesn't know what to do with herself.

She confides in me. What am I going to do *Hula*! I don't think I'll make it as a boat mate. Now I know exactly what that sailors' adage means: cruising is hours of pure boredom interspersed with moments of sheer terror. I've ended up bitchy and I hate that!

She got up abruptly and said I'm going ashore, Woof, but you've gotta stay here. She put some food in his bowl, grabbed her pocketbook and left. I saw her tying *Lilo* up near the bridge. Was she running away?

I was very upset of course. I love her and I love Jack and I love them being together. And they really love each other, I know. She can't leave!

I didn't have to worry very long. She was back in a few hours all bubbly and bright again.

She has rented a car AND she has made a deal with the syndicate that provides columns for small newspapers like the weekly she worked for in Connecticut. She phoned a guy who remembered her work at the *Crier* and he loved the proposal she just made.

She will write articles about the islands we visit. She won't have deadlines because there's no way of knowing when we'll be where or how long we'll stay. They're willing to keep it loose. She'll be paid $50 per article to start, more if the column is picked up by more papers.

Wow! Jack is so surprised! He had no idea that Jill wasn't 100 percent content.

She's really enthusiastic. I'll get out and meet the locals and learn all about the places we visit she told Jack. Plus I can contribute a little to the kitty again. That's what they call the pickle jar where they stash their cash.

Jill kept the car for a few days, driving all over the island, interviewing government officials and local fishermen, coming home full of facts and character sketches. She's perky and cute, like she was when we first met in St. Croix.

I'm not just Jack's girlfriend anymore *Hula* she exulted. I needed to have my own identity again! She quickly added Don't worry. I love being Jack's girlfriend. I'm determined to be good at that too.

I'm relieved, and I know Jack is too. Luckily he's

interested in all the information she brings back and can't wait to read her stories. Now he's teasing her about her outings and they're all lovey-dovey again.

The catamaran Jack worked on ashore has now been launched. It was a big day in the lagoon, with yachties showing up from all over the island to help ease her into the water—and enjoy the partying that followed. *Sirena* is her name. Maybe I should explain that all vessels are referred to as She and most boats have feminine names. Don't ask me why.

Sirena's a beauty and she'll be leaving for the Bahamas as soon as she's rigged and fitted with an engine. The owners and another couple have arrived. They're thrilled. They'll do a little island hopping on their way north.

Jack will be ready to move on once they're gone so Jill is busy writing her first article below decks, typing on the salon table where she's got maps and notes spread all over.

She read her lead paragraph to us.

This is an island with a split personality. It's small (34 square miles) but it's a possession of two countries: Holland, which spells its half Sint Maarten and where the inhabitants speak Dutch, and France, which calls its half Saint Martin and the language is French. Don't worry! English is the common language and you can drive back and forth without border stops and both sides love the American dollar.

J&J are sure this will be a popular column. This is going to be fun!

Oh! We have a refrigerator! A surprise present from Jack to Jill. He ordered it from Holland when he saw it in one of the catalogues in Hans' shop. It's about two cubic feet, about the same size as the stove, and Jack installed it next to the stove. He just bought another propane bottle to keep both appliances going.

Jill is thrilled of course. She ran ashore to buy a few perishables right away.

Antigua

We're in Antigua now. We're anchored in English Harbour, the home of Nelson's Dockyard, and Jill says she could devote a whole column just to it—its history as the West Indies headquarters of the English fleet in colonial days and its current fame as the center of Antigua Sailing Week, a big week-long international regatta every April, with yachts coming from all over the world to race or watch others race.

Today's yachties can thank Admiral Horatio Nelson for the excellent facilities ashore. Most of the colonial buildings have been restored and are occupied by small businesses catering to yachties, cruisers like us as well as racers. It's almost like a small village Jill says. She's shopped at the small grocery store and made friends with a lady selling fresh fruit and vegetables outside the gate so doesn't need to go to town.

Jack's already made use of the sail repair shop and small chandlery (that's a store that sells nautical gear) and they both go in every day for free showers.

Jill's catching up on our washing as there's a courtyard where anyone can wash and dry laundry. I like the comaraderie there she said. There's good vibes between the local ladies with their wash tubs and the boat women with their buckets.

They love the Admirals Arms. Neither one of them has ever been to England but they're sure this is a typical English pub. They had a beer there one day and plan to splurge on dinner one night before we leave here.

Oh, we've gotta be careful about the Woofer. When Jack went ashore to clear customs, which he could do right there at the Dockyard, he took Woof with him. Luckily another yachtie saw them coming and said Don't let the authorities see the dog. There's been a rabies scare.

So Jack left Woof with the yachtie, Jim, who hid him in the chandlery while Jack checked in. We moved out of the main part of the harbor immediately and are in Freeman's Bay just a short dingy ride away. It's quiet and secluded and the water is cleaner and there are nice places to take Woof ashore without being caught!

We're going to sail around the island. It's only 54 miles in circumference and there are 365 beaches to check out en route, one for each day of the year. That's great material for a column says Jill.

She's collecting material for her next article. She's already got a lot from the Dockyard. She intends to mention slavery.

Uh-oh, said Jack. Aren't you writing for people who just want a nice getaway vacation? he asked.

No, not at all she replied. They can read the ads for that. I think the people who come to the islands should appreciate the locals, be aware that almost all of them are descendants of slaves. I want them to know about the middle passage. I didn't know about that until I was living in St. Croix.

Jack said what do you mean, middle passage?

See? Jill said, You don't know either! That was the trip across the Atlantic from Africa to the Caribbean, in ships that piled captured men and women and even children into the holds, like cargo! They were chained! They couldn't stand up! They suffered for weeks like that! Then as soon as they got to an island they were sold as slaves! And treated worse than animals!

Well yes, I knew all that, Tweet. I just hadn't heard the term middle passage before. Why are you so angry?

I dunno she answered, suddenly deflated. I didn't mean to yell at you. I've just been thinking about slavery and how so many white people still treat black people poorly, like inferiors. Not you Sweetie. You're color-blind. That's one of the first things I loved about you. You appreciate everyone.

Jill was all teary and snuggled up to Jack. They just sat on deck quietly for a while, then Jack said Let's have a cuppa tea. That's what they've been doing for a break ever since Mystery Writer Mary came into our lives. They hugged and Jill perked up and minutes later they were sipping tea.

So, about the article, Jack said. Do you think bringing up slavery will go over with the editors?

Oh don't worry she said. I won't rant. But I think this is the island to mention it as part of the history. Antigua was the center of British shipping in the New World in those days so it had to be the center of the slave trade. I will also write that these descendants are now running their own country and doing quite well.

That sounds good but I have a suggestion. Don't send the Antigua article right away? Get your readers hooked on the series first with more light stuff?

Jill looked ready for a negative retort.

Hey, he grinned, don't forget I'm counting on you to support us! Some editors might not feel comfortable mentioning slavery. I don't want you to lose the job before you've hardly started!

Okay, she smiled. You've got a point. I shouldn't make it too serious, at least right away. Speaking of money, let's go to the post office to see if I've got a check.

They're having current mail sent to Yacht *Hula Hula* c/o General Delivery, Antigua, British West Indies. They haven't had any since leaving St. Martin a couple of months ago.

The Netherlands Antilles

We had stopped at a few other islands on our way to Antiqua. Just short stays. The most interesting were the two that along with Sint Marten make up the northern half of the Netherlands Antilles: Saba and Statia.

I've already told you a little about Saba. Statia (formally St Eustasius) is also not much more than a volcanic rock. Both, though currently Dutch, have mixed European backgrounds and small populations of mostly European-heritage people living there now.

(The southern half of the Netherland Antilles is

the ABC Islands off the coast of Venezuela—Aruba, Bonaire and Curacao. They will probably make another article if we ever get there. There are so many other islands in between! The Caribbean could take years!)

We were in Saba less than 24 hours, because there was no place to anchor! Here's Jill's column:

Small Specks in the Sea
Saba

Saba is truly a storybook island. Only five square miles, it is nothing but a volcanic rock shooting straight up from deep sea. There is no coastline or harbor.

Arriving by boat, we found a tiny indentation in the rock with a few fishing boats and a small cargo sloop tied up to some rickety docks. We rafted up alongside the sloop.

I guess it was quite an event to have a pleasure boat sail in. A young man from the power plant welcomed us. He had seen us coming and drove down the mountain to offer a tour in his vehicle, one of the few on the island.

There is one narrow road, built over a period of 20 years in the middle of this century, that winds up the mountainside. It connects five tiny villages, which are occupied by a total of 1200 inhabitants who live in small neat and tidy houses. Everything looks miniature, a bit Liliputian. Even the electric poles are short.

And, to complete the tinys, there's a tiny landing strip carved into the side of the volcano where tiny planes can land and quickly brake.

The people produce two products—rum and lace, lovely material that is hand-stitched by the local women in—you guessed it—tiny tiny stitches.

Sint Eustatius

Statia (formally Sint Eustatius) is slightly bigger than Saba—7 square miles—and has a little bit of coast. The only water shallow enough for us to anchor in was just off a beach. Our only neighbor was a small guest house and bar, The Old Gin Mill, which turned out to be a friendly hangout and great source of information. The "town," a small village with a fort, was at the top of a cliff above us.

We had hoped to explore, especially the crater at the top of the volcano where, oddly enough, there's a rain forest! And they grow bananas there! But we were unable to do that because there wasn't an available car anywhere on the island!

But we did manage by chance to be in the right place at the right time. This little Dutch island is, of all things, celebrating the 4th of July, U.S. Independence Day. How come?

It seems that during the American Revolution Statia was hopping as a smuggling haven. And as a politically neutral free port, it was an ideal site for transferring guns and gun powder from European to American ships. The supply was credited with helping the colonies win independence from England.

The new nation formally acknowledged the assistance of the smuggling island a few months after the Declaration of Independence was signed. Then an American ship, flying a flag never before seen outside the 13 colonies, sailed into Eustasius waters and exchanged gun "salutes" with Fort Oranje (on the cliff below which we were anchored) and the American captain presented the Eustatian governor with the first copy ever printed of that famous document.

AND, that's not all: In 1939 President Roosevelt visited the island to honor them again, this time presenting a plaque commemorating the site of "The First Salute". It says "Here the sovereignty of the United States of America was first formally acknowledged to a national vessel by a foreign official."

There's not much left of that era but the fort on the hill and part of the massive seawall built to create a bustling port. That was mostly destroyed by gunfire from British ships, enraged at the island's role in their defeat.

Statia is just a sleepy little island now but, we were told, it might soon be bustling again. Both The Common Market and Hess Oil are expected to build there. Don't ask me why. Someone said it's because of the deep water right up to the shore. That means that big ships, even tankers, can come right in.

Mail from Mary

We're in Martinique now. Jill is happy to hear people speaking French and to eat French cuisine. I think she could live on baguettes, croissants and Brie cheese.

She's commented that the local women, both West Indians and ex-pats, are almost as chic as Parisian women. (It's odd but she seems to love everything about France but the people. She likes their chicness and loves their cuisine but not the people themselves, resenting their disdain when during her junior year abroad she was constantly mocked for her American accent.)

J&J went to the post office first thing, hoping for more checks and Wow, Jill's ecstatic. She got a raise! She's now getting $100 per article because a lot more papers have picked up the column! The Antigua piece got special mention from her contact at the syndicate. The editors seem to like the idea that an island story doesn't have to be just about beaches.

So Jill hopes to dig deeper everywhere we go now, and Jack is happy that she's getting recognition again. She had been moping a bit about losing her identity, hearing herself referred to only as Jack's girlfriend or the girl on *Hula Hula*.

Now she's having her old editor at the *County*

Crier send her copies of the printed articles and she's not shy about showing them off. And the other night at a bar she was introduced as The Writer. So now she's happy again.

The other great item in the new mail was from Mystery Mary. It's the book! Jill shouted. Ohmygod Jack look she cried. The cover of *The Case of the Drowned Yachtie* is basically a photo that Jack took the day we sailed to Norman Island. They've added a sloop anchored offshore.

Mary's inscription inside is just as exciting: To Jill, the best typist/copy editor I ever had and To Jack, my favorite captain. Looking forward to more adventures together. What's more she gave them both credit in the Acknowledgements, Jill for her editorial assistance and Jack for the photo used on the cover.

How lovely said Jill. What do you think she has in mind asked Jack.

They found out in the letter enclosed. She'd like to get material for another Caribbean mystery while we're here. She was thinking that something involving a Trinidad carnival might be good and were we by any chance thinking of attending the next one and if so would we consider taking her aboard for a couple of weeks then.

It so happens we have been thinking about it. That carnival is supposed to be one of the best in the world, right up there with Rio and New Orleans.

Oh what do you think, Jill asked. I would love to have her. I think we could make her comfortable. She knows the boat so knows we're not luxury class.

I dunno he answered. I'm going to want to be with Trini a lot. (That's his pal from the AC crew in Tortola who is from Trinidad and is planning to meet us there. He goes home for carnival every year and comes back totally wasted.)

He's got Jack talked into doing all the pre-carnival stuff with him. Jack says that will be great article material for Jill. It's supposed to be crazy great he says—competitions for calypso singers, king and queen and costumes. The costumes are supposed to be fantastic. Huge elaborate affairs they spend a year making! But it all sounds pretty raunchy and boozy to me he said. I don't think it's quite Mary's cuppa tea.

Jill isn't so sure. I'm pretty sure Mary has a naughty side to her she says, and it will all make great material for her book too.

They left it that Jill would write Mary and explain that J&J will be participating in the raunchy events and if that's all right with her she's welcome to stay with us. The main carnival events are the Monday and Tuesday before Ash Wednesday but the partying starts a few days before that.

Carnival in Trinidad

It was OK with Mary but she almost didn't make it here. She was coming from England and flights from everywhere were booked solid for weeks around Carnival.

But she was able to pull some strings and get a first-class seat, arriving in plenty of time to witness everything.

They were all full of wows about the costumes after they went to the choosing of carnival king and queen. A real art form said Mary.

The main event for Trini—and therefore our crew— was *J'ouvert* (Creole for the early morning tramp from the outskirts of Port of Spain).

Swilling rum from a goatskin is apparently a must part of the tradition. Trini found enough goat skins for each of them, filled them with the local Caroni rum, and the group set off well before dawn to join the rest of the revelers for the rest of the day. They did it all—*J'ouvert* and the main parade that went on for hours. They were pretty sloshed when they rammed the boarding ladder with the dingy and staggered aboard that night.

J&J were obviously hungover the next morning but Mary bounced up the ladder onto the deck and announced she had her Trinidad plot! The murder would take place during *J'ouvert*. The victim would die of poisoned rum in his goat skin. She had no idea who the victim would be or who the murderer was or what the motive might be but she had her nugget, she said. She could spend the next couple of years thinking about that book.

Meanwhile she has already started one based in Singapore and has agreed with her publisher to place the next one in India.

You're covering the whole British Empire said Jack.

We call it Commonwealth now said Mary. Samantha likes to travel but she's not good at languages so I can't send her just anywhere! It has to be an English-speaking country. (Samantha, in case you forgot, is the amateur detective who solves all the murders in Mary's books.) I have to sprinkle them all over the globe, Mary continued, otherwise I might be tempted to follow your itinerary! Don't worry, she addressed Jack's wary look. You provide excellent material and I love being with you two but I promise not to be a pest. I don't suppose you'll be in Singapore or India very soon so it will be at least a couple of years probably before you see me again.

Meanwhile I'm having to make do with inferior helpers. You were the best assistant I ever had, Jill, and I wish I could take you with me.

Jill's Secrets

Jill has admitted, but to me and the Woofer only, that she does think about leaving. Not because she wants to work for Mary and certainly not because she's losing interest in Jack. I just don't know if I can keep doing this she confided after a few more rough passages.

The Anegada Passage was the worst possible way to start her cruising career but that was soon followed by a gale between Antigua and Guadeloupe when the main halyard broke. That's the wiry rope that holds the mainsail up! A major catastrophe!

Jack's immediate reaction was to climb the mast but thank god Jill screamed Get the harness! and he dropped back to the deck. I was bouncing around so much he might never have held on.

But the main still had to come down so first he dropped the jib, the only other sail still up, and that settled our motion a bit though I was still lurching around and the mast was still making great heaving motions and the mainsail was flapping all over and Jill was sure he'd be killed when he finally made the climb fully harnessed and managed to successfully drop the main and descend unscathed.

J&J bundled up the sail enough to throw it below, then kinda collapsed on the helm seat. Jill was still shaking and hanging on tight to unflappable Jack, who admitted he hoped he never had to do that again.

When the gale subsided we were thankfully near Guadeloupe and limped into the nearest shelter under just the jib and motor. It turned out to be a lovely big bay with a fishing village, *Deshaies.*

It was a nice place to recover from our latest disaster. Jack spent a few days repairing the sail and the halyard and when he went ashore hoping to find a part needed for the halyard Jill let me know she'd been scared to death and didn't know if she could handle it any more.

She tries to hide her fear from Jack because she's afraid he'll dump her as inadequate crew. I doubt that very much. He loves her so much and really doesn't need her as deck crew. He handles the sails himself and makes all the decisions and as long as she can take the wheel and follow his instructions in a crisis, and stands her watches on long passages, she earns her keep.

The good news is that after a few wonderful days at sea or ashore she's able to forget the bad stuff and really enjoy everything about the cruising life and the two of them have a great time together.

When they both went ashore to check out the village she was enchanted by the beach lined with beautiful fishing boats, all well cared for and prettily painted in bright colors—mostly blue or green outside and red inside. There was also an abundance of large fish traps created in a local "factory".

Farther down the coast we stopped at another gorgeous big bay—*Anse de Pigeon*—and saw another glorious sight early one morning. Two huge nets were being hauled up on the beach by a line of men and women dressed in colorful clothes, swaying in slow motion. Oh I wish I had a movie camera said Jill. That's a classic West Indian scene!

I love to see them bathing. If it's a nice sunny day at sea they drop a bucket over the side and splash each other with salt water on deck. If we're in a clean anchorage at a safe distance from spectators they jump over the side and frolic in soapy suds there.

They rinse off on deck with a tiny bit of fresh water. That's the most valuable commodity on a boat. Because it weighs a lot and takes up a lot of space it is carried in limited amounts and therefore used with utmost discretion. Refilling the water jugs is the one must-do item on every trip ashore.

When we are at anchor a playful bath overboard is almost always followed by a tumble on the bunk below. The Woof and I love to hear that.

The Venezuelan Coast

Since we left Trinidad we've covered a lot of ground. A quick trip to the town of Carupano on the mainland of Venezuela, just long enough to find the port captain's office and officially enter the country, then a few tiny islands off the coast of Venezuela, and then all three of the ABC islands, also known as the Netherland Antilles: Aruba, Bonaire and Curacao.

The Venezuelan islands were a welcome change from the big city of Port of Spain. As devotees of Jimmy Buffet J&J had thought they'd want to "do" Margarita Island but after all the partying in Trinidad they just wanted to get to the small mostly uninhabited islands. Of course we couldn't just sail by the big one. J&J must have spent an hour singing about "Wastin' away in Margaritaville" as we bobbed along that island's shore.

They always hope they might bump into Jimmy somewhere. He's been cruising the Caribbean and his boat was sighted near Tortola once when we were there. But so far they haven't lucked into a meeting. They dream of attending a concert. He apparently hangs out near St. Barts a lot but we never heard of a concert near there when we were in sailing distance.

We went on to the nice quiet small islands north of Venezuela and enjoyed a week or so just cruising from one uninhabited island to another, occasionally sharing anchorages with a small fishing boat or two but basically having them to ourselves. We caught a few fish for dinners.

Jill got out her typewriter and wrote a piece on Trini. Neither Jack nor I had any idea this was coming! Neither did Jill it turns out. But, she says, the more I got to know him and found out about him the more I thought wow, this is the epitome of the island boy. Just before he left to go back to Tortola I asked if it was okay to write about him and he said sure, just make sure to say I'm handsome! Here is what she came up with.

Island Boy

Carnival in Trinidad is every bit as fabulous as advertised. Its extravagant troupes parading through the streets of Port of Spain, the elaborate costumes that take a year to create and can take up the whole road, the eye-popping expanses of bare skin, the loud persistent calypso tunes pouring out of steel pans and sweaty exuberant bands on floats and the need for everyone—participants and onlookers alike—to dance along, to boogie, to "wine"—it's definitely contagious.

Trini is the nickname of the island and is also the

nickname of the friend who was our perfect guide—a 24-year-old native who is a good buddy of my captain Jack. They worked together on the boats at Antilles Charter in Tortola, where Trini is spending a few years accumulating a nest egg before returning home. He comes home for Carnival every year.

Carnival, by the way, is often called "Mas" in the West Indies, for masquerade.

I happen to admire Trini immensely. He is extremely handsome, his athletic body is a rich chocolate brown, his topaz eyes seem to offer everyone a warm welcome and his full mouth is usually curved in a beautiful smile.

He could be the poster child for a Trinidad citizen. He carries genes from every type of ancestor on the island—the African slave, the indentured Indians and Chinese, the European plantation owners and even the fierce pre-Columbian Carib Indians.

Trini's's favorite part of Carnival is J'ouvert, the early-morning tramp in which spectators, accompanied by multiple Calypso bands, sing and dance and prance their way through the city to the center of town where the formal Carnival will soon start. It's a bacchanal in which the locals really let go. Almost everyone carries a bottle or flask or goat skin of rum.

Trini pointed out a few startling "marchers," angry looking men covered in oil or mud or paint, and explained they show up to make a statement. They still dwell on their legacy of slavery and don't want anyone to forget it. But they eventually got caught up in the melee too and there was never any trouble.

J'ouvert participants can't be confused with the official revelers who later appear with their various troupes on the streets of the town center in gorgeous and elaborate costumes, many so huge they are supported on wheels. These revelers also dance to lively Calypso tunes from multiple and thrilling steel pan bands, but their moves might be a tad more sedate.

Trini lured us into the parade to momentarily follow his favorite floupe (steel pan band on a moving truck). We danced along in close contact and got a good look at 'wining' in action. This is a very suggestive move similar to what we call bump-and-grind!

It was definitely the highlight for the Hula Hula crew, but we enjoyed every bit of Carnival thanks to our Island Boy guide.

A day or so after we all recovered from Carnival, Trini took us to his multi-generational family home in the country, where we got to admire the well-tended farm and the warm-hearted family of our friend and to better understand the source of his sweet, lovable personality.

We were warmly welcomed by four generations enjoying a Sunday afternoon of rest. We ate mountains of strange-to-us but delicious foods; drank a number of different homemade beverages, some alcoholic, all sweet and delicious;

and "limed" with several outgoing characters of all ages. To lime is to hang out with in the islands.

The interaction of family members was so good-natured and respectful I wish I could have filmed it as a how-to for American families. I now understand better the deep respect islanders have for their ancestors, and I no longer wonder how our friend Trini got to be so special.

The North Coast of South America

We did the ABCs quite quickly, just a few days in Aruba and Bonaire and only one in Curacao, not a place to visit by boat we found. To get into the harbor, where we had to clear customs in the big town of Willemsted, first we had to wait for a pontoon bridge to open. We circled and circled until finally a pilot boat came along and the bridge opened for it.

Once inside it was late afternoon and there was no place to anchor so we tied up alongside a Venezuelan fishing boat, which was very welcoming. Every time the cook cooked something he sent a sample down to J&J, from a cup of coffee to dinner and as a parting gift, five little fish. Jack gave him a bottle of scotch.

We spent just a few hours. J&J "entered" (did the formalities) and took a quick walk along the main street. Jill loved the pastel colors of the buildings which she thought looked very Dutch, although of course she's never been there.

They bought a few essential groceries and got back aboard. They thought we might find a nice anchorage along the coast but nothing appealed so we took off for Aruba.

Jill had another I-can't-do-this-anymore afternoon when we sailed over a bank just off the coast of Aruba. The swells were monstrous and on the beam (hitting us sideways). One actually lifted my port hull out of the water. We made it through the swells and found a quiet anchorage off the airport. It was close enough to dingy or ride the bike into Oranjestad, the town, and Jill said she felt like Alice in Wonderland when she found "a gorgeous supermarket" and spent $55! She also found a bikini sale and bought 3 of them for $5 each.

The shopping spree was followed by two rainy days so Jill was again ecstatic that she could catch up on laundry. Every pail got a workout and one of the coolers used to store rice and crackers etcetera was emptied temporarily for use as a rain catcher.

It's a good thing it doesn't take much to make Jill happy. She seems to love each island and is easy to please. Jack lets her run the show when we're at anchor. I know he'll do anything to keep her spirits up. Nothing will keep him from sailing around the world but he'd be lost without her.

Jamaica

From Aruba to Jamaica was a long haul but a very pleasant sail, with the wind abeam (that's on the side) most of the time. We were able to keep up the main and the genoa all the way. Jill got to enjoy a long stretch without one catastrophe. I'm delighted that she got to see how pleasant cruising can be for more than a few hours at a time.

Our first port of call in Jamaica was Port Antonio, on the north coast, and that's where we spent most of our time. It's a nice little harbor that we had to ourselves. There was only one other vessel of any kind, pleasure or commercial—a small launch that plied across the harbor from the mainland to Navy Island, a tiny islet once owned by the American movie star Errol Flynn, now being done up as a resort with cute little thatch-roofed cottages.

J&J like it here and plan to stay a while. Jack says it would be hard to beat as a place to work (on me!) and a place to leave me for a while. They want to take a trip inland. This is the biggest island we've ever been to so there's lots to see.

Port Antonio is our kinda town, Jill says.

Everything's within walking distance, and the people are friendly or shy. It's a very tumble down place where life seems to have stood still for the last several years. Nobody is fixing anything up.

There is a huge abandoned United Fruit wharf but bananas are no longer shipped from here. We were told that there are usually one or two cruise ships in the harbor but so far we haven't seen one.

The market is marvelous. Lovely fruits and veggies and low prices. Everything grows here, including marijuana. Young guys approach Jack at every step.

Jack started a couple of projects right away. He moved the mizzen mast forward a tad, AND he's wiring me all over for fluorescent lights! I think Jack would have been happy with the kerosene lamps forever but when a buddy in Tortola gave him a fuse box salvaged from a VW wreck he invested in three fluorescent lamps! Now he's finally installing them. Another gift for his Tweetie! She's thrilled of course. There will be one in the galley, one in the salon and one in the center of the starboard cabin, where the head and tiny basin and the tiny mirror are.

Then they were ready to start their inland travels. They took a train over the mountain to the capitol city of Kingston, which is on the south shore. That meant going up and over a mountain.

The story I heard when they got back! They had neglected to factor in that it was Saturday, market day. The little diesel train had two cars. It departed at 5:45 a.m. and was scheduled to arrive at 9:15.

As the train climbed the mountain it stopped at every opening in the trees to board men and women loaded down with stalks of bananas and baskets of fruits and vegetables for the market, even a few chickens and goats. Somehow they kept packing them in. The seats were overcrowded in no time and the aisles were so crowded the conductor couldn't get through. Jack was squeezed into a middle seat and took a lot of shoving around.

Jill luckily was next to an open window and enjoyed her view, seeing so many people living off the land. Sometimes she could see into their tiny houses but mostly there was no sign of life until the train stopped and the market-goers appeared out of the trees.

They had expected to spend a couple of days in Kingston but quickly changed their minds. Once they peeled out of the train, they set off walking, expecting to explore on foot but they were soon dissuaded from that by various citizens who approached them and cautioned them not to go down this road or venture into that neighborhood. It seems that white visitors were being badly treated in many sections of the city.

A friendly taxi driver approached them and offered to give them a tour at a reasonable price, which they accepted. They saw a museum, the university, a large art gallery and a popular hotel where they stopped for a meal and decided to go back to Port Antonio that day.

They found an open bus scheduled to make that trip and boarded. It seemed quite picturesque at first but as the bus kept stopping to board more passengers (now it was the farmers returning from the market) it became less so and soon they were as squashed as on the train, hanging on for dear life as the little bus navigated the bumpy roads.

It was definitely not equipped for this mountain pass. They had to get out and walk up some of the steepest inclines, then hang on tight when the bus careened down others. J&J weren't sure the brakes were working at all! They were a total mess when they fell aboard that night.

Jill had named this The Honeymoon Cruise as it was the first serious cruising they had ever done together. Hoping to make it more romantic for her Jack set off for Montego Bay on the eastern end of the island, a renowned honeymoon destination.

Predictably neither one of them liked the fancy hotels and they were very turned off by the shops in town with their tacky tourist wares and aggressive sales people.

So our Caribbean cruise is almost over. We're going back to Tortola where Jack will spend one more season as a charter captain and they will get me ship-shape for the beginning of our round-the-world journey.

On our way home we do not plan to stop. We will sail past the eastern tip of Cuba, where Fidel Castro doesn't want us, and Haiti, where Papa Doc and the Tonton Macoutes wield tyrannical terror over its people and visiting yachts are required to berth in marinas with a local guard aboard. We'll also bypass the Dominican Republic and Puerto Rico because we're just in a hurry to get "home."

Where's Woof?

We got a huge welcome home from the guys at Antilles Charter and the bar flies at the Bilge Bar, where the cast of characters hasn't changed that much. J&J really relaxed for a few days and had a ball catching up with the gang, including Trini. A few of the old friends have sailed on but a bunch of new yachties have sailed in.

All was great until one of the old pals asked Where's Woofer? Oh dear. I haven't mentioned him lately because Jill falls apart whenever she thinks of him.

When we were in Trinidad we left him with Trini's family because we knew we would have to leave him behind some time before we left for the Pacific and it would be hard to find a better home. We had already had a taste of the British quarantine laws and had learned that in the Pacific they are very seriously enforced. There a dog must stay on the boat for six months before being allowed ashore. There was no way J&J would subject the Woof to that, and Trini's extended family and their extensive property seemed perfect for his new home.

Woof loved running around the fields and playing with the other animals and being petted by numerous children and elders. J&J are sure there couldn't be a better home for that dog, though Jill really mourns his loss.

I miss him too but I'm glad that since he couldn't come with us they found him such a good home. And frankly, he'll be much happier there. He really wasn't all that keen on sailing. He dove down the hatch as soon as the motion got rocky, and he did a lot of shaking under the table in the far corner of the salon. He wouldn't come back on deck until he heard the anchor chain rattle. Then he bounded up the steps and ran to the port bow, ecstatically sniffing land.

He's a dear dog and I'm sure he's grateful that Jill rescued him and J&J gave him a home and family but I'm also sure that he's relieved to be living ashore with a loving family in a place he can lead a good dog's life.

I feel sorry for Jill. She was so attached to that dog. But she knows it has to be and will manage.

The Five Month Plan

This will be an interesting winter. Both sets of parents are coming for two-week visits, their last chance to see us in the Caribbean before we take off for the circumnavigation. Who knows how many years that will take? Jack reckons maybe 10.

J&J know one family that spent only three years. But they had three teen-age kids aboard. The daughter jumped ship in Australia (later marrying the guy who caused the jump and producing five children) and the two boys were eager to go to college back home.

Our scenario will be quite different. The cyclone season is six months long, a good time to stop sailing and find a nice place to anchor and find jobs. Jack's already got the first couple of years figured out.

In the South Pacific, where we're starting,

cyclone season is the opposite of ours, so Jack intends to cross the Panama Canal in June and cruise through the several Polynesian island chains until December, when we'll head for New Zealand to sit out the storm season. We'll then cruise through Melanesia and Micronesia for the next six months before holing up for another storm season in Australia.

They both intend to find jobs during extended shore times. Both sets of parents have offered to help them financially but they think that working their way around the world is part of the point of cruising! They'll get to know the people and soak up the culture of the place! I like that!

Jill has been much more into the whole idea since she learned that we'd be more or less stationary for six months of every year and she could fly back to the States to visit her family each year.

But first we have to get started. We've got five months to get me ship-shape for travel to the Panama Canal.

I've gotta interrupt here to explain what seems like a major problem to many people. Why are we not going to Hawaii? Jack builds a Hawaiian boat and gives it a Hawaiian name but you're not even going there?

I gotta say I think that's a very good question but Jack and most other sailors know the answer. You wanna go with the wind. And the wind below the equator is better for sailing west than the wind above.

Jack already has several charters booked for January through March. He won't take any more on because he needs plenty of time to devote to me. It seems there are still lots of important accessories I need to take on the big wide world.

Jill has made up her own list of things to do before we set sail again. Lots of sewing and painting. She loves making me pretty. She's got cute little curtains over the bunk portholes already and intends to put them everywhere. She makes sure Jack OKs her fabric choices. They've gotta be sorta nautical.

Preparations

In between skippering jobs my captain reverts to his other persona—Jack of all trades. He has a long list of must-dos before we take off for the Pacific.

First is a self-steering vane. So far Jack and Jill have taken turns at the wheel every time we went to sea but with the long passages coming up they know that non-human help is essential.

A self-steering device (basically a wind vane connected to the wheel) is a pretty new concept and the few that have been manufactured are quite expensive. And by nature unpredictable. Obviously each boat reacts differently to each type of equipment. With all that in mind, Jack is making his own—of course.

He sits on the deck cutting very thin plywood into more or less rectangular shapes. These he "sews" together with fishing line. With every gust of wind he holds the contraption up to see what happens, then tweaks things a bit.

They've already christened the unwieldy gadget George, after a gangly awkward friend.

Then there's fishing gear. He's bought plenty of heavy duty line because in the middle of the big ocean he assumes we're going to be catching some very big fish. He's got an extensive collection of lures—from screw drivers to feathers—everything he's ever been told never fails.

I gotta say I haven't seen much fishing luck so far, but then we haven't tried very hard. We've never been so far from a grocery store that we absolutely had to catch a fish! But now we'll be trawling for days or weeks at a time and a freshly-caught tuna or wahoo will be very welcome.

We'll be trawling two very long lines. Each will be attached to a specially-repurposed winch that will be secured to the stern crossbar. Another plus for a catamaran: Thanks to the broader width the boat can easily drag two lines without worrying about them getting entangled.

I probably need to translate: a winch is a cylinder attached to the deck with a handle protruding from its top. When sailing, a sailor winds a winch handle to tighten a sheet. A sheet is a rope attached to the corner of a sail for that purpose. We have one of these winches on each side for both the main and the mizzen.

The Parents are contributing to our equipment too. Theirs are more tangible forms of survival. Jill's parents have already sent a life raft, a round rubber double tube with an international orange canopy that compresses into a "valise" that is stowed on deck.

Jack's parents will bring an EPIRB (Emergency Positioning Indicator Radio Beacon) with their luggage when they come next month. This is a gadget the size of a fire extinguisher that sends a distress signal across the ocean up to 150 miles away. It will be stowed on deck with the life raft.

Of course it's kinda scary to plan for all the things that could go wrong. I can see Jill crawl into herself whenever a disaster possibility is the subject of conversation. She has been reading too many survival stories! The one about 100 days in a life raft almost made her pack her bags.

And it didn't help when she asked Jack if they should practice man-overboard rescue techniques. His response was don't even think about it! With only two of them aboard and an emergency most likely to happen in rough weather, only a miracle would save the one in the water.

He explained that it would be highly unlikely for the one remaining on board to be able to keep his eye on the one in the water, toss a life ring, turn me around and ....STOP! Jill shouted. Okay okay! Scratch man overboard training. I don't even want to think about it! But we've gotta do whatever we can to keep ourselves on deck. I'll be a total wreck if I have to worry about you going overboard whenever it's rough.

We'll be fine Jack replied. First of all look at *Hula Hula* here. She's designed for rough weather! She's basically upright most of the time. If it's rough and you have to be at the wheel you'll be right here on center deck, as far as possible from the water. If I have to move around, go forward or aft or to the side, I'll wear a life vest and a harness. I'll rig it with an attachment—a line attached to a big strong clip so wherever I have to work I can clip myself to a lifeline or whatever else is handy. Okay?

Jill looked worn out, as if she had just been through a storm, but gave a weak okay. Now she's got me worried too. There's no way one would continue without the other. That EPIRB doesn't reassure me. The Pacific is so vast the chances of a rescue at sea seem pretty slim.

The Sanfords

Jill's parents are here! They're terrific people, even though they know nothing about boats and have never been sailing. And they're not going to start now because Cathy, the mother, gets seasick.

They're staying ashore at the Fort Burt Hotel in Roadtown, a nice old inn perched on the cliff overlooking the town and the harbor. Jill's rented a car so she can show them the island and be with them most of the time.

Of course they were eager to meet Jack and me so they came aboard their first day. They admitted they had their doubts about Jill's choice of a new life but once they met Jack and me we won them over.

Cathy and Bob Sanford had never been aboard a catamaran before—or any sailboat for that matter. Jack gave them a terrific tour, explaining just enough of the nautical stuff so they could get a sense of what was involved in cruising and playing up the wonderful job Jill had done to make me so attractive. They love my color scheme, the turquoise and teal.

Jill had made brownies—her specialty—and tea was served. The Parents already knew from her letters that Jill had gone British since being in the

BVI so long and were delighted to share a cuppa. Cathy was also very pleased that the sewing skills she had taught her daughter were being put to great use.

Our center deck is our living/dining area. It has a solid plywood "floor." Jack has attached slatted wood settees to the sides of the cabins to port and starboard of the deck to serve as sofas. Jill has covered foam cushions with a nice navy blue naugahyde to rest on top of the settees and she covered bed pillows with subtle print toweling covers as back rests. With the nice low folding table Jack made for the center of the area we now have a handsome comfy home.

The Parents liked all that a lot but they especially admired the awning and spray dodger J&J made together. I better explain. The canvas awning is our "roof." It's about 20 foot square, as it spans my 20 foot width and the middle of my length. It protects us from sun and rain when we're at anchor and thanks to funnels sewn in at strategic points and plastic pipes to carry the rainwater into jerry jugs, it provides us with our main source of fresh water.

The dodger is also used only at anchor. Like the top of a convertible car, it is canvas stretched over curved metal pipes. It's 10 feet wide, attached at the bottom to the forward end of my center deck. The top can be raised into a curve shape about 3 feet high to protect us from wind and/or horizontal rain and any spray that might jump up through the great big nets that are slung between the hulls fore and aft. The Sanfords couldn't believe that J&J made those huge complicated items themselves, especially once they got a look at our sewing machine.

I may not have mentioned that our antique machine does not have a treadle. Because of course we don't have power. So J&J sit cross-legged in the middle of the deck with the machine on the low table. Jack somehow manages to wrestle huge swaths of canvas into the eight-inch hole and crank the wheel while Jill uses both hands to feed the material through the hole, under the needle and out!!

The Sanfords knew enough about sewing to be totally flabbergasted. Bob endeared himself to me when he declared By god if these kids can do that then sure, no problem, of course they can sail around the world!

The visit's going well. I think it hasn't been mentioned before but there's a great coincidence in the J&J story—that both fathers own hardware stores, which means that both are great handymen.

Bob has been a big help working on the starboard cabin. Up to now J&J had nowhere to stow their clothes. They just piled everything they own onto the forward bunk. His and hers are all mixed up.

So the guys are building a cedar "closet" and two lockers, one on each side of the head. (A reminder if needed: the head is the toilet.)

The lockers (really just cubbyholes) are about two cubic feet each, nicely tucked into the swoop of my hull. They have hinged covers. I know Jill will be making cushions to make comfy seats for them.

It now looks very ship-shape down there. Jill immediately wanted to see if her clothes fit in. She gave away a lot when she moved aboard—the nice clothes she had worn to work and on dates. Her current wardrobe of shirts and shorts and jeans and bikinis fit in just fine. Needless to say Jack's collection of rags leave plenty of air room in his locker.

The closet is across the cabin between the ladder and the tiny basin. It protrudes just a few inches. What do they need to hang up? Jill has two slinky long party dresses she's determined to keep and Jack has one sport jacket. He didn't even have that until he got dragged to a dinner party in a fancy restaurant. They'll also use the closet for shoes. Although they don't wear any aboard, and they wear just flip-flops ashore, they each have a pair of real shoes they need to keep dry.

On the side of the closet next to the ladder are two hooks, one for each oily, now easily accessible from on deck just by reaching down the hatch. Oily, by the way, is Brit slang for oilskin, which is Brit for foul weather gear, which is basically a hyped up slicker used by most sailors. J&J have not invested in oily trousers, figuring that since they aim to stay mostly in tropical latitudes they won't be needed. I hope they're right.

The final addition to the cabin is a book shelf. There's a little space for it above the head, under the two port holes. J&J are both big readers and this is for their leisure-time reading, if they ever get that. Jack's nautical books—on celestial navigation, etc.—are tucked into a shelf in the saloon, where the charts and sextant are.

Provisioning

While the guys are smarting me up, the gals are in St. Thomas shopping. J&J have made up a list of stuff we need to have on board when we set sail for the Pacific, mostly a lot of non-perishable food that we'll need for long passages. Cathy offered to help her daughter stock up and Jill thought that was a great idea. They will make a day of it—lunch, points of interest, duty-free shopping on Main Street.

Jill carefully chose the date of the excursion on a day that no cruise ships were scheduled to be in port.

St. Thomas is a popular cruise destination and the big harbor can accommodate several cruise ships at a time. But the town of Charlotte Amalie is small and when thousands of passengers overwhelm it Jill doesn't want to be there. So early this morning the gals boarded the *Native Son*, a ferry that makes frequent runs between Roadtown and Charlotte Amalie, the town in St. T.

The plan is to meet up with Barbara, a friend of Jill's from St. Croix who is now living in St. Thomas, and fill her car. She will spend the day with them, have lunch with them and get them and all their goodies back to a late ferry. But before that Jill hopes they have plenty of time to do Main Street, a shopping stretch famous for a vast assortment of duty-free goods, from jewelry to cameras.

As for the provisions, they will probably find everything they want at one big supermarket, including booze! Which is duty free in the USVI!!

Stowing the Goods

The gals arrived home with so much stuff! The ferry staff wasn't fazed by it but when Jack saw the boxes coming off the boat he said whoa. We can't possibly fit all that in *Hula*.

We'll find a way Jill said and darned if she wasn't right. About a dozen boxes are now empty, their contents stowed away in every nook and cranny I have.

The bunk forward of the galley has been saved just for food. Dozens of cans of vegetables and meats and soups and dozens of bottles of booze, all carefully wrapped, are now tightly packed into the V under the bunk boards. Would you believe vodka for $1.25 or a bottle of Cruzan Rum (it's made in St. Croix) for 99 cents??

On top of the boards are a couple of good-size coolers for storing things that won't survive long in dampness like flour, sugar, rice and pasta. Everything they're likely to want in the near future is carefully packed on the galley end of the bunk. Jill is sure she doesn't want to crawl into that bunk when we're underway.

More Articles

Once her parents left, Jill decided she should write a few more articles before we start cruising. She doesn't want to lose her readers' interest when we're incommunicado on long passages. And, of course, to make a little more money.

She's just tossed off one on Puerto Rico and one on the USVIs. Her friend Cookie is Puerto Rican and when Jill lived in St. Croix she often joined Cookie for weekends in San Juan. She obviously got to know the big island quite well. And of course while she lived in St. Croix for a few years she also got to know St. Thomas and St. John fairly well so was able to write up the USVIs.

Puerto Rico

For most Americans Puerto Rico is the most familiar Caribbean island, thanks to its large segment of population that has relocated to the States.

Like the U.S. Virgin Islands, its inhabitants are citizens of the USA but do not have full rights. They cannot vote for president and have no representation in Congress except for one "delegate" who can serve on committees but cannot vote. (For some reason they have different tags. Puerto Rico is called a commonwealth while the USVI is called a territory.)

Unlike most of the islands I have written about for this column, Puerto Rico is a big one, one of the Greater Antilles. Its population hovers around three million.

Thanks to its mostly Spanish heritage, Spanish is the common language, though most of its people speak English too, especially those connected to the tourist industry, which is thriving.

You will know you're in for a charming experience when your plane touches down at the San Juan airport. The minute the wheels touch the tarmac the Puerto Ricans aboard clap their hands (and often follow that up with the sign of the cross).

San Juan, the capitol, is the hub for tourists. Many large hotels line the shoreline, their pristine beaches the center of activity every day, their elegant casinos and nightclubs beckoning every evening.

At least one day trip into Old San Juan is a must. A port city, the remains of two forts book-end the waterfront. The vast harbor behind the town is bustling with cargo ships, cruise ships and pleasure boats.

Built on a steep hillside in the 15th century, most of the charming old buildings have been beautifully restored as private homes, shops of every description, restaurants, art galleries or other businesses. Narrow cobblestone streets tie it all together.

Night life in the old city ranges from flamenco dancing at a hotel in a renovated convent to a drag show in a sleazy waterfront bar.

For those who want to venture outside the metropolitan area, the island has many wonderful surprises. Here are a few of my favorite things.

The interior is hilly and mostly agricultural. Its inhabitants are fondly called jibaros, which I interpret as country bumpkins. I strongly urge finding an excuse to interact with them as you drive through their lovely countryside.

In fact, you should make a point of getting to know the people all over the island. In the States they are often looked down upon as "others" and "immigrants". In their own environment you will find that they are a happy, kind, fun-loving people. There is usually music in the air. And often dancing.

The second largest city is Ponce, on the south shore. There the Museo de Arte de Ponce makes the one and a half hour drive from San Juan a must for any serious art lover. A surprisingly large and excellent collection of paintings by renowned European artists dominates, though Puerto Rican artists through the ages are of course given equal space.

The building itself is a work of art, designed by the American architect Edward Durrell Stone, who also designed MOMA in New York and the Kennedy Center in Washington.

Even if surfing is not your thing you'll probably be glad you made a visit to Rincon, a world-famous surfing destination on the northwest corner of the island. There the reliable surf is fantastic to watch roll in, and chances are you'll see a few seasoned surfers ride the enormous waves. It's a feat to thrill most spectators.

The most fabulous find on Puerto Rico is the Arecibo Observatory on the western end of the north shore. A massive federal project, it is the world's largest radio telescope: a

1,000-foot round reflector dish in a sinkhole on the ground and, cable-mounted 500 feet above, a steerable receiver and several radar transmitters.

This behemoth of an instrument attracts scientists from all over the world. They gather radio signals from stars, planets, distant galaxies and whatever else might be out there.

The U.S. Virgin Islands
St. Thomas, St. Croix, St. John

This trio of gorgeous tropical islands, billed as The American Paradise, can please almost any vacation-goer. Their beaches are world-famous and so is their duty-free status.

If you're wanting to get away from it all, St. John is your spot. Sparsely populated, it is mostly National Park and appeals to nature lovers. It's the only island with a campground.

If you want some action and/or love bargains, St. Thomas is what you're looking for. It's famous for its extensive duty-free shopping and jumping night life in the town of Charlotte Amalie. It's also one of the most popular cruise-ship destinations in the world.

St. Croix is uniquely un-touristy, a plus to many travelers. Separated from its sister islands by 40 miles, it is larger and flatter and its history as "the garden spot of the Caribbean" during colonial days is apparent.

A drive through the countryside reveals the previous existence of dozens of plantations, mostly which had prospered immensely in colonial days by growing sugar cane.

Sprinkled among the ruins are some beautifully restored plantation greathouses, where the European planters lived, and the remains of about 100 "sugar mills", wind mills in which the sugar cane was crushed by African slaves.

St. Croix remained rural in subsequent years, until the 1960s, when Hess Oil built the largest oil refinery in the western hemisphere on the south shore of this quiet island. Alcoa Aluminum soon followed with an aluminum refinery next door.

Luckily St. Croix is big enough that most residents and visitors can overlook the existence of heavy industry and simply enjoy the natural beauty of the rest of the island and its many plusses, including the third largest barrier reef in the hemisphere, a bonanza for divers, and a lovely historic botanical garden.

It's an island that has escaped too much development and thereby saved much of its natural resources. Outstanding is Salt River Bay, the only uninterrupted "web of life" left in the Virgin Islands if not the entire Caribbean. Bird and fish life abounds, as water flows from the top of Blue Mountain into the river then into the bay, where it nourishes extensive mangrove forests and seagrass beds that feed large populations of birds, both locals and migrants, and a wide spectrum of sea life, then out to a coral reef and beyond to a submarine canyon.

There are two towns. Christiansted is laid-back and slightly sophisticated at the same time, with some elegant shops featuring European china and crystal side by side with casual or gourmet outdoor restaurants. There is almost always some soft jazz in the air. Frederiksted is a bit of a ghost town, but a friendly one.

If natural beauty, the fact that no building is taller than a palm tree and a casual lifestyle appeal to you, St. Croix is your island in the USVI.

It's a place so friendly and charming I chose to live there for a few years, before I met the man and the boat that whisked me off to sea.

The Lindsays Are Here

Jack's parents are here. Ed and Kelly Lindsay. The nicest thing is that they also own a hardware store so Jill feels right at home with them. But they're very different from her parents because these are sailors and know all about boats. They've done a lot of sailing off the New England coast but never sailed elsewhere before. They are enjoying their first sails on a catamaran immensely and they love the BVIs!

J&J are happy about that because they love to show me off and they're delighted to have a last look at these fabulous islands with such appreciative passengers aboard.

We're checking in at most of the favorite islands. Cooper, where J&J have become good friends with the British couple running the restaurant there. Norman, where we showed them the setting for Mary's next mystery story. It turns out Kelly is a fan of Mary's! And of course Gorda Sound. Where of course they were blown away by the Baths.

So much so that we anchored off them a few days while Jack made use of his dad's assistance with myriad last minute tweaks to sails and rigging, compass and steering vane, etc. Everyone swam ashore at the end of the day for another wander through the magical forest of smooth giant granite boulders.

Jack admitted to his parents that they made him miss sculpting. Oh good his mother said. I hoped you might get back to that now that *Hula* is finished. Jack surprised everyone by saying I dream of a quiet evening somewhere like New Zealand when I can sit down and whittle.

I can't wait said Jill. You've done nothing but create a boat since I've known you. Not that I consider that a mere nothing but ever since you told me you once did some sculpture and showed me the picture of that dog table I've been dying to see your more delicate artistic side. You've really buried that!

Kelly chimed in good girl Jill. Make sure he makes time for sculpture. He really has talent.

And Ed added since you insist on working your way around the world and won't accept money from us maybe you can make money sculpting. I've been worried about the jobs you're likely to pick up as an itinerant. It's nice to think you might be carving instead of pumping gas!

Bravo! said Jill. I'm going to make sure he makes time to carve.

Ed insisted on checking Jack's tool and spare parts supplies and immediately made a long list of things he'd be sending immediately from his hardware store.

The four of them had some fun evenings sharing memories of kid adventures in the magic world of hardware stores.

Here's the piece Jill wrote just before we finally sailed away from the BVI.

The British Virgin Islands

This cluster of 60 islands northeast of the American Virgin Islands is best known for two wildly different reasons. It's a major tax haven and it's a sailors' paradise.

It's also a delightful vacation destination. The main island of the group is the biggest, Tortola, which is all of 36 square miles. Some of the others are inhabited, some not. Some have tourist amenities, others not even a hut.

It's ideal territory for exploring by boat and sailing is its chief attraction. It is becoming a prime center for the relatively new charter boat business. You can hire a boat for a week and sail it yourself or be provided with an experienced captain. Most of the boats can accommodate four or six people, and the provisions provided include the best wines and steaks.

If you're into racing the Tortola Spring Regatta is a highlight of the Caribbean racing season. Yachts come from all over the world to participate.

Two things I recommend you don't miss:

The Baths: This is a gorgeous grouping of huge granite boulders on a stretch of beach in Gorda Sound. They are up to 40 feet tall. They are smooth and shapely and are gracefully draped all over each other creating arches and tunnels and grottoes and tidal pools. It is breathtaking and awe-inspiring to wander through.

Foxy's Bar in Jost Van Dyke: Foxy is a local Calypsonian with a marvelous ability to make up lyrics on the spot. If he happens to take a fancy to you you might hear something about yourself in his next song.

Goodbye St. Croix

We're back in St. Croix where we started, where I was created! What a welcome we got.

We're kinda famous here. Not many boats have been built here by amateurs and certainly no other boat snatched away a reporter from the *Island Times!* So the editor sent Jill's replacement over to interview us about what we've been up to the last couple of years and what's in the plans. J&J spent an hour giving her a good story.

Part of the good story is that we're here not only to say goodbye to old friends but to borrow the vice president of Christiansted Bank. Mike, Cookie's boyfriend, is going with us for the first long haul--through the Panama Canal and on through French Polynesia! He's taken an indefinite leave of absence from his job. He's thrilled about going to sea again. He was in the Navy in the Vietnam War.

That's a big relief to me. J&J have never been at sea for more than a couple of days at a time! We might be at sea for a month! Who knows how they'll handle it! I'm so glad to know they'll have help!

Cookie and Jack's old roommate Ben are planning a big bash for a going-away party. We are back at the dock where I was put together and there will be a big barbecue on Ben's lawn here. The theme will be South Sea Islands. Everyone's getting out their sarongs. (Here in the VIs everyone has Java Wraps thanks to a local resident who imports them from Indonesia). And Cookie is somehow rustling up leis.

We'll see Cookie again in a few months. The plan is for her to fly to Tahiti when we get there. Then she and Mike will cruise through the rest of French Polynesia with us and probably fly back from Bora Bora.

Jill finally gave up the lease on her apartment

here. Jack teased her for keeping it so long and put on a big show about how worried he was that she had planned to back out at the last minute. Jill let him know that many times that hadn't been too far from the truth.

I gotta admit that I've really worried about that. So it was a big relief when the subject came up and Jill made it clear there was no way she was letting him go without her. She gave him a bear hug and a big smoochy kiss and said I may be crazy but I'm not letting you go without me.

Oh, beautiful! What a relief that was!

Jill thinks we'll be back. She thinks this is where we'll settle down after the circumnavigation. Jack's not so sure. He's open to whatever happens.

Panama

We're in Panama! Waiting our turn to cross the Panama Canal! Then we'll be in the Pacific!! Jack has been waiting for this for years!

We came here almost directly from St. Croix, more or less a straight shot southwest. We had large swells most of the way, some roller coasters as Jack calls them, but most of them luckily were going our way so just helped push us along. We mostly bounced and lurched along a course of 240 for exactly a week, either running or on a broad reach, often just under the jib.

At one point no one had touched the sails for two days. Jill said it feels like The Flying Dutchman.

We made only one stop on the way, just off the coast of the Panama mainland. That was in the San Blas Islands, a large group of about 370 small islets off the northeast coast.

We had barely dropped anchor off Porvenir, the main island, when a canoe with two women came alongside. They were dressed in wonderfully decorated colorful clothing that Jill knew all about.

They're selling molas she cried. I've been dying to see them. I'm going to buy one and Mike you've gotta buy one for Cookie too. She will love it.

The canoe was loaded with unique panels of vividly colored cotton appliqued with multi-colored geometic designs or figures of flora and fauna, mainly birds and fish of all descriptions.

The women were invited to come on board but declined, instead passing a large basket full of the cloth up to Jill on deck.

Jill had done some homework and told the boys the inhabitants of these islands were of the indigenous Kuna Indian tribe. Because they were physically quite distant from the Panama mainland they had managed to keep their tribal ways virtually intact.

Their custom of painting their bodies with pictures of the natural world had segued into wearing clothing adorned with such pictures. And with the advent of international tourism these molas had been discovered by the rest of the world.

While the crew looked through the goods on deck the women happily hung on to my bulwarks, bobbing up and down and smiling and giggling as the crew oohed and aahed over the contents.

The panels were in table mat size but many were also sewn into hand bags and blouses. In fact they realized the two women were wearing such blouses and Mike said aha I see the one I want for Cookie—that one, indicating the one on the older woman.

Giggle giggle. No problem. After a short exchange in pidgin/English/Spanish it was decided that when our crew went ashore the next day they would meet up with her and pick up her freshly-washed blouse.

Meanwhile Jill was so smitten she bought many more than one placemat. She was sure her mother would love them.

Ashore the next day they found that these islanders had been able to keep many of their customs intact. Most of the inhabitants had probably never been off their little island.

Here's what Jill wrote in her log:

They live in tiny reed or thatch huts. All the women and girl children dress as if for a festival every day. Their mola blouses are connected to wrap-around skirts that are usually of some loud print that has nothing to do with the colors or designs of the molas. Most of them wear a gold ring in their noses and a black stripe painted down their foreheads and great clunky necklaces of gold or silver coins or other ornaments and lots of small strings of beads of all colors wrapped around their arms and legs.

This is the only place we've been in the Caribbean where there's quite a culture visible to the most casual observer.

She'd love to write a column about it but isn't going to because tourists are unlikely to visit them. If they go to Panama they'll buy molas on the mainland and are unlikely to take a boat trip so far offshore.

When we pushed on to Panama proper we had our first encounter with a large ship. There are hordes of them heading for the Panama Canal. Unfortunately Jill was on watch in the middle of the night when it loomed into sight. She watched it for a while until she realized the space between us was decreasing and she quickly woke Jack, who leapt onto deck and immediately spun the wheel. The ship was soon well out of range and we resumed our course.

Jack is so sweet about training Jill. She was

blaming herself for letting the ship get too close.

No, my fault he said. I never thought to talk about the possibilities. How were you to know! Actually the rule of the sea is that sail has the right of way but don't believe it. I doubt most ships pay that much attention. You did the right thing to wake me when you did but now our rule will be to wake me the moment you see a ship anywhere within a few miles. I don't want you to have to make decisions. I'll sleep better if you promise to do that Tweet. I'll never be too tired. And the same goes for bad weather. If you see black clouds looking too threatening don't wait to wake me. I'll want to be on deck to figure out what to do.

He gave her a big hug and sent her below. He and Mike stayed on deck the rest of the night. Ship traffic got thicker and thicker as we neared Cristobal-Colon, the city on the Atlantic side of the canal. The anchorage was packed with ships inside and outside the breakwater.

We made our way to the yacht club, where we anchored in "the flats" for a week while preparing for the big transit! We were warned about going into Colon. One yachtie we met had been mugged twice and another warned be prepared to fight! So both our guys made just one trip in together for absolute essentials, leaving Jill safe at home.

The Panama Canal

Our transit was unique. Most yachts tie up to the side of a ship for the 12-hour passage but Jack talked the authorities into letting us go "center lock" because he worried that my fragile plywood bulwarks would be shredded rubbing against a big steel ship. That meant we were alone in the middle of this wide wide canal, with four 120-foot lines connecting us to four men, two on each side of the canal, one forward and one aft, who walked us along the at least 10-foot high walls on either side of each lock as it filled with water. These four guys were our "mules", replacing the little trains that pull the big ships through.

We were required to have a "pilot" on board. Mr. Ripley was ours. He sat on deck quietly until we got to Gatun Lake in the middle of the crossing and Jack started raising sail.

No no he shouted. You motor.

No no said Jack. It's at least two hours to cross and I'm not going to putt-putt across if I can sail. It's a beautiful day. You'll enjoy it! And he did.

On the other side of the lake we picked up new handlers and continued on through more locks. All uneventful until the last one, when the gate opened to the Pacific Ocean and a terrific current sent me whipping around the lock and our handlers let go of the ropes!

Ohmygod this is it I was sure. But Jack gunned the engine and headed for one wall, then at the last minute reversed and headed for the opposite wall, back and forth, back and forth until finally we were headed straight ahead and shot out through the gate. What a way to arrive in the Pacific!

How does he always know the right thing to do! I think we all thought this was the end, including the pilot, who sat with his head in his hands during the whole ordeal. When we pulled into the little station where he was to disembark he could hardly walk.

Someone untied our four long ropes and everyone marveled how with those swirling around us we had managed to avoid fouling the rudders and/or the prop.

But cool Jack didn't seem overly shaken up like the rest of us. It was dusk and we motored over to a nearby sandbar, slipped in behind it and anchored and the crew crashed.

Then we spent a week in Balboa, the city on the Pacific side of Panama, where my crew stocked up on plenty of fresh food for the long haul. Jack figured our crossing to the Marquesa Islands would take 30 to 40 days.

Then we found a beach where I was hauled out to have my bottom cleaned.

I don't think I've mentioned that one of the great benefits of multihulls is that we can be beached for cleaning and painting the bottoms of the hulls. That's a chore that single-hull owners spend a ton of money paying for at a boatyard where a lift is required to suspend the boat while a crew takes care of the hull.

In my case we spent two days on a nice beach, where my crew scraped and cleaned my bottoms (the underwater parts of my hulls) during one low tide and painted me during the next. Now we don't have to worry about gathering a garden on the long cross. Then we were off.

THE PACIFIC

Jill is keeping quite a detailed log for the big crossing, our most historic event so far, so I am letting her tell the story. Her dear little typewriter shares space on the salon table with all the navigation papers and equipment. Here are her logs.

Panama to Marquesa Islands
35 Days

Day 2

It was a much less dramatic leave-taking than St. Croix.

No friends to wave us off, no sleepless nights in anticipation, almost no butterflies. I seem to have gotten all the fear out of my system

We more or less drifted out of Panama yesterday but by 5 the wind came up and the sea stayed flat and we really took off. Very exhilarating.

The boys wanted to keep both jibs up and the pole out all night to make the best of it while it lasted so I was excused from watch for the first time. Lovely, but I was awake most of the night anyway because it was kind of scary going so fast and slewing around. It felt out of control from below but wasn't once I came on deck.

Probably the main reason for the lack of excitement about leaving Panama is that we've decided to go to Isla de Cocos first, which is just a 500-mile trip as opposed to 4000. It's a tiny Costa Rican island with no settlement but on the way and a place to break up the trip and maybe top up our water and get coconuts and lemons. Don't swim is the message everywhere around here, and we've seen a few sharks already. Last night a spectacular porpoise show. Several of them darting around the bows for an hour or more, all covered in phosphorescence.

Day 4

We really lost the wind after the first day. We've been sailing in fits and starts since then. After 152 miles the first day it was 31 miles Saturday and 75 Sunday. The sea has been lovely and flat the whole time, which pleases me. There are porpoises and sharks and birds all over. Yesterday afternoon we saw a huge school of porpoises, dozens of them, as far as you could see, really racing toward the southeast up and down, up and down, probably as fast as they could go.

We're having a fruit orgy. Bought a bag of 100 oranges before we left and have had to throw out a lot already. So I'm making juice every day plus we're eating lots of them. Plus we had a papaya already and another one is ready and we're working on a pineapple. When those few delicacies are finished it will be apples and oranges and cabbages and onions forever.

Day 5—We're in the doldrums. Absolutely no wind some of the time, just a breath other times, now and then enough to actually sail for a few hours. We've run the engine twice now for a couple of hours, once to get away from some thunder storms.

We've been going five days and have only 375 miles behind us. (It's some 4000 to the Marquesas!) This is supposed to be the worst area for wind. We certainly hope it improves. It's very hot and sweaty. I still have to wear a shirt as I get burned even through the big Martini umbrella Jack scored for the helm seat.

We see the occasional ship, most look like fishing boats. Lots of fish today and we just caught our first Pacific fish, a baby dorado. We have our order in for another as it's hardly enough for a bite apiece.

Day 6 &7: Cocos Island.

This is an uninhabited island except for several men at a National Park station. They were very hospitable to us, helped us fill our water jugs. Jack was able to reciprocate by fixing a ham radio for them. They said we could send some mail. It'll be picked up by a bi-weekly patrol boat and stamped from the mainland.

Then we had a great rain so spent hours washing ourselves and everything on the boat, all our clothes and sheets and towels.

There are supposed to be a lot of sharks here so we're not swimming but we haven't seen any. We were told that Jacques Cousteau came here and saw more varieties of sharks than anywhere else in the world—something like 20.

Day 8

We took off yesterday late p.m., found a good east wind which we hadn't expected and no fish, which we had. It's continued to blow fairly well for 24 hours. We can't quite make our course—southwest—but we have plenty of time to correct that.

We're all remarkably calm about setting to sea for 30 or 40 days. This is really it but we're in such a routine already we're really casual. No momentous thoughts at all. I've got to say I'm glad I didn't back out. So far anyway.

Day 11

Our fourth day out of Cocos and we've been very lucky

with wind. We've done over 400 miles. To be sure we're tacking south, but at least we're moving in this area of usual calms. Only 150 miles to the equator. We have a bottle of champagne chilling for the occasion. Strangely enough it's rather cool here. We sleep under a blanket. Two big currents meet here, the warmer one from the north and the cooler one from the south.

We should see the Galapagos Islands tomorrow. We'll not be stopping there because we'd be charged a lot and have to take a guide aboard. I just read Darwin's report from his voyage on the Beagle and what really impressed him was that in addition to the many species not known elsewhere, each of the islands had different species from the others.

The islands are volcanic and dry and barren on the lower 1000 feet or so. The upper parts are always in the clouds and are lovely and green and fertile.

Not much happening. We eat, sleep, read and navigate. Mike and I take noon sights. He works them out and logs the official reading. I'm still practicing using the sextant. Jack takes star sights morning and night when conditions permit and other sun sights during the day. Mike is about to start

Advanced Nav-- longitudinal sights of sun and stars.

We had one rainy night and Mike got the worst of it. Three hours of steering because George was stuck. We have to steer now and then but mostly George does it or we can lock the wheel.

No fish. We don't even see them anymore. No sharks or porpoises since Cocos but plenty of birds.

<u>Day 12</u>

We are becalmed off the Galapagos Islands—Culpepper actually, just a big cake-shaped rock. Beyond it is another really strange rock. It looks like the Arc de Triomphe. Really!

There are porpoises frolicking about and I saw my first whale fairly close—swimming along and spouting, then sounding with a big flourish of tail. Dozens of birds. They circle the boat like they're playing a game. They sit in the water as we bear down on them and don't move till the last minute.

I harvested a bumper crop of alfafa and mung beans in the sprouters Cookie gave me. They're marvelous! The best going-away present ever! I was afraid they'd use up too much water but they're doing fine with very little. Jack calls the sprouts my babies.

Our water capacity, incidentally, is 85 gallons. We try to limit ourselves to a half gallon per person per day and we do pretty well at it. We use salt water for cooking when we can, for brushing teeth and washing bodies with just a squirt of fresh-water rinse.

We've been eating very well. We still have quite a lot of fresh things. Not a lot of variety but luckily we all like oranges and coleslaw. Favorite dinners are corned beef hash and chili, straight out of the cans. Spaghetti, fried Spam and mashed potatoes (out of a box), macaroni and cheese. I'm using corned beef in spaghetti sauce and chili and it tastes fine. Mike is really into bread making so we have sandwiches every noon. And I've invented a popular house juice: a weak combination of orange and lemon Tang that's good with our drinks. We allow ourselves two drinks a night.

Day 13

Such a busy time. We didn't lose sight of Culpepper till 2 this afternoon. We spent a lot of time becalmed and just barely moving but we couldn't have picked a more interesting spot to spend some time. We had a seal playing around the boat last night. A big fat guy with long whiskers who played a lot like a porpoise and made the same kind of blowing noise only louder. He'd swim up to the side, then dive under and play between the hulls. At the same time there was a bunch of porpoises circling us, one guy jumping straight out of the water and falling back with a loud smack.

We finally got some wind this morning and as we got going we had a big escort of birds. Mainly cormorants, we think they are, very curious and unafraid. They circle the boat very closely, always looking as if they're going to poke their bills through the jib. They crane their necks to get a better look. One perched on our starboard bow and stayed there for so long we thought we had a figurehead for the crossing and named him Pepper in honor of Culpepper Island. He let me get very close to take a Polaroid shot and looked at everything with great interest. But he left before we could find him a mate for the other bow, which we would have named Salt of course.

Shortly after we got started we saw what we thought was another yacht on the horizon. Then we realized it was something much smaller and closer. It was a bamboo raft with a wigwam-type structure supporting palm fronds—for sails? For shelter? Were they survivors of a wreck? Refugees? Were there dead or dying bodies aboard? Skeletons? Finally we were right there and it was moored. The palms were a birds' nest and lots of birds were aboard. A hand-painted sign said something about Ecuador and anchor and we figured it was some kind of scientific experiment.

So much excitement for one day!

Oh, then a terrible thing. One of the cormorants took our fishing lure and was drowned by the time we pulled the line in. We were trolling a silver spoon.

Now we just saw three whales spouting to port.

What next?

Unless we get becalmed again we should cross the equator tomorrow and if we're already in the trades which seems possible, we just turn right and go straight ahead for the next 3000 miles. The first 1000 are supposed to be the hardest. They've been slow—13 days—but if that's hard I'll take more of that.

<u>Day 19</u>

Yesterday we crossed the equator—twice. What happened was we were quite sure we'd cross it about noon and were all set to have our champagne for lunch, then Mike worked out the noon sight and we were already across, by 20 miles, more than we expected but oh well hooray--pop, bubble and fizz. At 5 o'clock Mike took another position sight and uh-oh, things weren't working out right. Ultimately they figured out we had been 20 miles <u>north</u> of the equator, not south. We were really crossing it just now!

I'm experimenting with baking in the pressure cooker. I tried a big cookie last night (which I had never even done in the conventional way) and it came out looking like a fall-apart cake, of course, but tasted delicious.

Porpoises and whales everywhere! How wonderful it would be to have a really good movie camera on a trip like this!

Jack and I have read Melville's "Typee". It's about his four months living among the natives of Nuku Hiva (our destined landfall), when they were still cannibals and in other ways unspoiled. What an idyllic place it was. You just reached up into a tree to get your food, which was a good thing because the natives were so "indolent" they wouldn't have been very good at cultivation. They were all happy and carefree and healthy and friendly (except to enemies).

Day 21

Only 2,500 miles to go. It's amazingly cool here on the equator. The sun is hot but if there's any breeze at all it's cool. We have a sheet and maybe a light blanket on the bunk, day or night. Frequently it's too cool to sit under the umbrella. And we went for two days without baths because it was too cold. The water is too.

We think we're in the tradewinds now. Yesterday the wind came around to the southeast and has been blowing steadily ever since. We've been doing six or seven knots. The wind and seas are on the beam so it's a bit bumpy but we all welcome a little speed for a change.

Day 22

We're really barreling along now, doing about 7 knots, making up for all the slow days. We'll be at the half way point in a few days and are planning our half way party. The wine

Mike bought in Panama, our only bottle aboard, and to go with it probably spaghetti. At that time we will make a pool on our arrival date and time in Nuku Hiva. Winner gets a free dinner ashore or something.

I've finally settled down to really doing the noon sights, i.e. working them out. Then I check against Mike's figures, as he's the official noon sighter. I think I've got it pretty well in my head now but am no way ready to go on to lines of position. They seem much more complicated.

The porpoises around here are smaller than we're used to and they have white snouts. We haven't caught one fish since Cocos.

Day 23

Today we have really "boisterous" seas for the first time. Hula is bouncing around and spray is flying and we're making seven knots.

Jack says I should include some of our menus, such as today's lunch, which was several slices of beer bread, Carr's table water crackers, caviar, salami, cheese whiz, sliced onions, mustard sauce, alfalfa sprouts, one slice of fried cold spam, peanut butter and jelly. Can't say we don't have variety. Last night we had fried spam and rice pilaf and coleslaw and the boys put leftover stew on their rice.

I had a wonderful watch last night. Three showers in three hours between midnight and 3 and I had to be on deck the whole time because I couldn't get George to steer itself. My red coat leaks and it soaked through Jack's jacket, my sweatshirt, a cotton shirt and corduroys and undies. Will have to get my old yellow jacket out.

Day 25

Last night was a fun half-way party. We dressed: I in a pareu, Jack in his party pants and a headscarf and machete, Mike in his striped pants and canvas hat. A wahine, a pirate and a trader, sort of. The wine was Greek and not wonderful and I woke up with a headache last night and have had it all day. But we had music and candlelight and fun.

We made our pool choices for land ho in the Marquesas. Jack and I picked May 17 (two weeks), I at 9AM, he at 15:31:22. Mike's got May 20 at 6:20.

Today we finally caught a fish. Unfortunately a tuna. A

beautiful blue, fat one but Mike and I don't really like them and Jack's not wild about them. But we'll give it a try.

Day 26

It was delicious. Mike has a way with fish. He marinated it in lime, sherry, vinegar, onion, oregano, salt and pepper and we all loved it. We saved what was left over and I made a salad of it. Threw it in a bowl with the other leftovers—rice and coleslaw—and that too was delicious.

I'm working out the sun sights now and making few mistakes. I'll have to move on to the next course soon. Also starting to learn a few stars as I sometimes do the paperwork for Jack's star sights.

I'm trimming the sails a bit these days too. Especially at night time when the wind is light and the wind shifts and it seems a shame to wake up Jack so I started fooling with the sheets and that was fine with him. Such a feeling of accomplishment to get the boat going again. It does make it more fun and interesting to participate more.

We've had some real wind the last couple of days. Last night we were averaging 9 knots. It was a little wild.

The time passes so quickly. This is our 24th day from Panama, 15th from Cocos, and no one seems the least bit stir crazy yet. We keep busy—cooking and cleaning in addition to sailing and Jack does odd little jobs now and then when it's calm. In between we do a lot of sleeping, reading and looking around.

Day 28

We're really burning up the miles. Jack has started keeping track of knots with his new calculator and 6.3 is slow. We could be there in a week.

We're growing mildew all over the boat. I'm cleaning a little every day.

We're having our last coleslaw today. I think the cabbages would have lasted even longer if they had not been refrigerated and in plastic bags when we bought them. The moisture stayed with them. We're down to a few oranges that are all dried out. Still have good onions and potatoes. I'm happy as long as I have onions and eggs for the rest of the trip. Eggs for baking more than anything, as I have to make cookies about every other day.

We make bread every other day too. Although our meals are necessarily simpler than usual, I still seem to spend most of my time in the galley. And yet I'm really being spoilt by the boys when it comes to washing up. One of them does it almost every meal.

We're playing Scrabble at cocktail hour and that's fun, even though I'm not winning.

Day 30

Less than 1000 miles to go. If we keep up at this rate we'll be there within a week. I remember the days when that many miles and days seemed a momentous undertaking and I didn't think I could do it. Now it's a mere nothing, just the last lap. We feel we're almost there.

We're now planning to go first to Hiva Oa, rather than Nuka Hiva, because we can enter there as well and it's downwind of NH. And it's near Fatu Hiva which is apparently the most beautiful of the Marquesas and has the best tapas, if not the only ones. So we'll be delayed getting our mail, but hope we'll be able to phone or cable from HO.

We're finally sailing with twin jibs, which we thought

we'd have all the way. Instead we had the wind on the side till a couple of days ago. We're really surfing most of the time, doing 8 and 9 knots. At night we take out the poled jib and slow down a bit.

I'm doing exercises every day. Am up to 40 sit-ups and 8 push-ups. I'm so busy with things I have to do and want to do that I sometimes get all upset about the time I have to give to the helm. Crazy.

Day 32

We caught a beautiful dorado today and are going to try to eat most of the meat one way or another. Mike is marinating enough for two meals. We'll chill half of it in the "letter box" (that's what they call the tiny ice compartment of the tiny fridge.) The rest Jack cut into thin strips and is trying to dry them on the stern. He's got a grillwork going with the ladder and some strips of wood.

It's almost impossible to find a dry spot on the boat right now as we're getting a lot of spray. It went all rough and horrible the night before last and though it's improved now we still have quite big seas and are taking them more on the beam because we've altered course to a bit more southerly.

I took my first position line sight today. Shot the sun around 9 a.m. and went over the sight form with Jack. About twice as much to it as the noon sight but I trust I'll be able to get it eventually.

The last two nights we've seen a loom on the horizon, looking like a ship with a lot of lights including a searchlight. Last night we got close enough to see that it was a fishing boat. Spooky to see that loom.

Yesterday when the seas were at their highest the steering line broke—the line that connects the wheel to the tillers. It had broken about a week ago and Jack replaced half of it then. This time he put a new line all over. But George was steering and the tillers were swinging wildly in the big seas and I got all worried about Jack being knocked off but he's very good at hanging on.

Day 34

We've got both jibs and the main up trying to keep up enough speed to get to Hiva Oa before dark tomorrow! Yippee! I'm getting excited. It will be nice to stop for a while. Although

I haven't been bored and actually enjoyed it most of the time it will still be nice to arrive. Mainly it will be nice to stand and walk again without wobbling. My balance has been worse than ever this trip. The following seas and the big swells I guess is the worst combination. Even Jack has a little trouble. I'm forever sitting down abruptly and walking on all fours.

I've taken noon sights and worked them out myself the past two days. Yesterday I had one mistake. Jack was very proud of my first effort. Today I think I got it perfect. I'm surprised at myself that I can do it at all and really surprised that I even enjoy doing it. I had looked at it as something I ought to do, never as something I'd like!

Day 35!!

We arrived about 5 PM, having sighted Hiva Oa at 11. We could see two other small islands as well. Great excitement. The coastline of Hiva Oa is all rocky cliffs and green green, like a moss or short grass growing all over. We headed for Atuona, which is the main town.

We anchored in a small bay with nine other yachts, including the orange catamaran we had met in Cristobal. We had too many drinks with them, followed by a wonderful full night's sleep.

THE MARQUESA ISLANDS

It's me again—*Hula Hula.*

I guess you can imagine what a relief it was for me to drop anchor in Hiva Oa! Can you imagine working around the clock for 34 days? Of course not! And doing it without ever complaining? No way! And with never making a mistake? That's right. I performed perfectly the whole voyage and I gotta tell you I deserve a good rest.

I'm happy to say J&J really appreciated me and have assured me they're not touching that anchor for at least a week. We all need to catch our breath.

We've spent a couple of weeks in the Marquesa Islands, one of the groups making up French Polynesia. (The others are the Tuamotos and the Society Islands.)

The Marquesas are Hiva Oa,Tahuata, Fatu Hiva, Nuku Hiva and UaPu. Quite a change from the Caribbean. Spectacular ragged volcanic peaks, shrouded in mist, rising from rolling green hills. Just like in National Geographic Mike said. Many beaches with black sand.

Most of the bays are quite small and rough. Going ashore the crew has to deal with surf, and laying at anchor is called rock 'n roll.

When an inter-island trader arrives it anchors out and the goods are transported in by smaller boats. As they near shore a couple of guys jump out, turn the boat into the wind and hang on tight as others, up to their waists in water, unload boxes and crates and walk them ashore on their shoulders or heads. Quite a production!

Our first anchorage was Baie Tahuaku on the island of Hiva Oa. We met Philippe, a young French single-hander, whose crossing in a tiny yellow boat less than 20 feet long had taken 79 days!

To "enter" all three of my people went ashore because each was required to pay a steep bond price, $850, the cost of a ticket back to their homeland, for the privilege of visiting French Polynesia. Jack figures there must have been a lot of yachties going native in an unattractive way for the government to seem so prepared to evict them.

The second most important task was trying to contact Jill's parents, who expected a call the minute we arrived. As non-sailors they worried a lot about their daughter going to sea and needed to be reassured frequently that she was okay. In the

Caribbean she called home every two weeks. Although the parents knew the crossing would take longer than that, Jill knew they would be a wreck after a month of silence so was quite desperate about making a call.

Bad news. For some reason there was no way to call America from Hiva Oa. But she could send a telegram, which she did. (When she finally was able to actually talk with her mother she learned the telegram didn't arrive till five days later!)

In general they are all impressed with how civilized the villages look, especially when compared to the Caribbean. Neat and tidy yards with stone walls or flowering hedges around them. Grass instead of sandy dirt, no garbage in the streets or yards.

Fatu Hiva was the chief destination in this island group because in yachtie lore it is the lushest, most fertile spot in all of the Pacific! Everyone was ready for a feast of fresh food, and we found it.

To get there we had to beat to windward for the first time since the Caribbean. I had almost forgotten what it felt like.

The crew was also eager to practice trading. Since the locals don't have much use for money, yachties exchange a variety of goods for the fabulous fruits grown here.

We anchored in the Baie des Vierges in the village of Hana Vave. J&J had read it is the most beautiful bay in the Pacific, and they are not disappointed. But none of my crew has managed to spy the virgins said to be found in the huge rock formations. Apparently the locals and most visitors see them, especially the one called Virgin and Child.

Not too disappointed, they set out armed with a lot of trading stuff and had a ball trading one T-shirt for a sack of the biggest pamplemousse (grapefruit) ever seen, another shirt for a big stalk of bananas, two combs and two barrettes for five ripe papayas and some oranges, and a bag full of limes for free. You help yourself by pulling them off the tree.

They also had a bunch of coconuts. We are learning there are many groves of coconut palms on most of the Pacific islands, left from the plantations of colonial days, and the locals harvest the copra (dried nuts) for traders of the soaps and other products made from dried coconuts, but there are so many ripe ones left lying on the ground that everyone just helps himself.

The other must-do in the Marquesas was to see tikis. These are very old wood carvings found in the bush. J&J made a trek up a hill on Nuku Hiva, where they found quite rudimentary carvings of several massive human forms strewn about on the ground—only one whole one standing.

They are glad they made the walk because they

passed several houses where large family groups were "lolling about" on verandas, according to Jill, and all the houses looked fairly substantial and well-cared for and had pretty flower gardens with huge tropical plants.

She also reported that the women were wearing bras with their pareus tied around their waists. Pareu, you've probably gathered, is French for sarong.

The people on all the Marquesa Islands are generally quite light skinned, a yellowish olive Jill calls it. Whatever indigenous people were here when the European colonists arrived were obviously outnumbered.

The Tuamotos

These are the atolls of French Polynesia. An atoll is a more or less round circle of tiny islands joined by coral reefs with a large lagoon inside the ring. I can tell you that it's no fun getting into or out of those lagoons. There are only tiny openings in the atolls and you've gotta go with the tide because the current is fierce.

Our first encounter was with Takaroa Atoll. Jack's pilot book described this opening as about 80 yards wide on the seaward side, much narrower at the lagoon entrance. It also reported currents of 5 to 6 knots and advised entering only at slack water.

We did not have a tide table so the crew had to judge by eye. It looked all right when we started in but when we got abreast of the quay where the village is the current caught us and Jack had to do one of his excellent but scary maneuvers to get us turned around and tied up to the dock. There were two other yachts there—a Dutch trimaran and a German sloop. A big crowd, what Jill said must have been the whole town, watched our performance.

We had missed the slack tide by two hours so would have to stay at the dock overnight and go into the lagoon the next afternoon.

J&J decided to check out the village. When it started to rain a man invited them into his house. He turned out to be the mayor and the island's best pearl diver. They had a good conversation in French, his not much better than theirs.

Among the things they learned: the village has 125 people, used to have much more but now most of the young people go to Tahiti. There are more Mormons than Catholics. No Protestants. The school goes to age 10 or 12. If a kid wants more education he goes to Tahiti. They decorate their homes with shells. Ropes of them hang from roofs. The land is divided among all the people so everyone is in the copra business.

When J&J commented about some huge pearl

shells displayed in his house they discovered he was a pearl diver, one of the best. He can dive 30 feet for them. They left with a big pile of shells and were so glad that it had rained and they had that great visit.

The lagoon was not what we expected. It's definitely not sleepy. It's three miles wide and so deep—10 to 15 fathoms except around the extreme edges—that you never see the bottom, just darkness. And we actually had surf. The flora ashore is jungle-dense—lots of pretty trees and bushes as well as palms.

There were six other yachts in this lagoon and my crew and a pair from another cat rowed ashore to see a wreck—County of Roxborough, a four-masted iron schooner that went aground in a hurricane in 1904. They were impressed that is was in quite good condition.

Then an overnight sail to Apataki. There are two passes into this lagoon and Jack chose the north pass because it had "only" a three-knot current but wow it was rough. The waves were like stone walls up to my beams, one after another every couple of feet. Luckily it was a short pass so we were through soon.

There was supposed to be an anchorage near a village but we sailed all over and didn't find one. Eventually we anchored behind a reef. The next morning the wind had died and the sun came out and

Jill said it's just like a lagoon should be.

The boys went snorkeling and saw a shark.

We moved on and finally found a village with a fish-processing plant. We tied up there. Lots of speed boats coming and going, all used for fishing there.

The crew took a walk through the tiny village and were again impressed to find it mostly neat and tidy. Friendly people, everyone saying *bonjour*. Two stores but everything closed on Saturday.

A bunch of kids hung around at night. Some are imported from Tahiti for fishing. They dove in the water, danced to Abba on a portable tape player, seemed high on something.

A gray cat came aboard and fell in love with Jack.

We left at dawn for Tahiti! And saw two whales, one very close! They were swimming toward us, then turned and swam alongside, one only a couple of feet away, the other about 100 feet out.

THE SOCIETY ISLANDS

Tahiti

Our arrival in Papeete, which is the capital of French Polynesia and the main town of Tahiti, was a bit chaotic. The harbor is quite small so yachts are not allowed to swing on anchors. Instead we're required to tie up stern-to the quay, which puts us side-on to the wind. But we finally got settled.

Mike rushed to a phone the minute we were settled to call Cookie and tell her we are finally here, to hop on the first plane she could get. She plans to cruise through the rest of the Society Islands with us, then she and Mike will both return to St. Croix.

Jill has been laid up for a few days with an infected cut on her leg. It started from a tiny scratch she got on the rusty wreck on Takaroa. It came to a head like a boil and she can hardly walk. Her treatment consists of hot water/Clorox compresses, antibiotic cream, bandages and antibiotic pills.

Like all good yachties J&J have a good medicine kit on board, omitting anything that might contain an illegal drug. In some countries the authorities conduct very thorough searches of arriving vessels. Even a drop of codeine in cough medicine could lead to an arrest. They also took a first aid course before they left Tortola. And they have a medical reference book.

So, while Jill is on sick leave—mostly reading on the bunk-- the boys are using the down time to pay attention to me! They're mending sails and frigging with the rigging and tinkering with the motor and fixing this and tweaking that.

I certainly needed a little attention. I mean really! When you think what I've been through the last few months! All those long passages and that horrible Panama Canal transit when I was almost shredded! And that endless VERY long crossing when I didn't rest for 35 days!!!

Surely I'm due for lots of attention! And affection!

Oh, I admit all three of the crew have toasted me now and then and Jack is always muttering good job *Hula* and Jill croons sweet nothings to me all the time and even Mike is impressed. This is one helluva ship he's said. She's taking a lot of punishment without any complaint.

I guess what I mean is that it's nice that Jack is working on me again. I like having his undivided attention and knowing that he knows he couldn't do it without me any more than I could do it without him. We are a team, essential to each other.

Cookie is Here!

I guess you can tell that we all love Cookie. We do! She's Mike's long term girlfriend, Jill's best friend ever, Jack's "favorite kook" and, for me, the only female other than Jill who's ever given me any attention! She has always given me a nice pat on the cabin top and said Hi *Hula* whenever she came aboard.

She thanked me for getting the crew here safely. She has an open-end return ticket so will be with us throughout the Society Islands.

Her arrival timing was good because by that time Jill could walk again. Cookie's a shopper, which Jill is usually not, but right away the two girls took off to the big town of Papeete.

The town got high points—attractive, neat and clean, friendly people. Not at all French, Jill had to add.

They praised the shops—a nice combination of chic French boutiques and local craft shops. Very nice shell jewelry. They even raved about grocery shopping, splurging on lamb chops, Chinese vegetables, brioche and Brie. And of course they both bought a couple of *pareus*, finding some with beautiful batik designs.

Most nights the crew goes out for dinner at "Les Trucks"—a parking lot taken over each evening by a bunch of trucks that serve food from counters along the side. The biggest bargain in town. Each meal, whether steak and chips, Chinese or crepes, is $5.

They set out for the yacht club one evening in the dinghy but on the way found a man with two cold kids and a half-sunken outrigger canoe on a reef. After towing them ashore it was too late to go to the yacht club so they tried again the next night and came back with another good story: an English yachtie named Sid went aground off the yacht club. Members towed him in, hauled out the boat, repaired his engines and sails and re-provisioned his boat! All for free!

During the days they did some touring of the island by bus. Of course they wanted to swim but all the beaches were uninviting—black sand and dirty brown water full of coral all over the island.

The Gaugin Museum was a must and another disappointment because there were no originals, only prints and those not very good ones.

Paul Gaugin, in case you don't know, was the French artist who lived many years in Tahiti near the end of his life and is most renowned for his paintings of the topless native women there.

It turns out he spent his very last years in Atuona on Hiva Oa and is buried there, the very village where we made our landfall. The crew can't believe they missed that landmark!

Moorea

Jill is in love with this island!

To quote a bit from her log: This is the true paradise. Gorgeous green hills and mountains plus beautiful beaches and a barrier reef that completely surrounds the island. The people are lovely and their homes are attractive, beautiful flowers and shrubs everywhere. I haven't seen one unattractive thing here. I could stay forever and be perfectly happy, I think.

That last line really scares me. Here we are just getting started on our circumnavigation and she's already thinking about where she could stay forever? Help!

We anchored in several different places around the island, dodging lots of coral heads but also finding some of the best snorkeling yet in gorgeous turquoise water.

Some bays have attractive little resorts, with cottages made of native materials—bamboo walls and palm thatch roofs. Nothing is over one storey high.

Maybe the highlight of this port of call was a lunch at Club Mediteranee, Cookie and Mike treating. All I know is that the crew gorged themselves on the huge buffet AND brought home a huge doggy bag with enough goodies for another two lunches and one dinner!

Huahine, Raitea and Tahaa

We didn't spend much time in these lesser islands because we were rushing to get to Bora Bora before Bastille Day. The crew had heard great things about the celebrations there and the days of *tamure* dancing competitions.

Huahine will be remembered for what Jill describes as the most beautiful swimming hole I've ever seen. This island is completely surrounded by a reef with a wide lagoon between it and the island. We anchored in the middle of the turquoise lagoon and Jill spent most of the day just swimming around or floating.

The girls did a little shopping in the small town and Jill came back to report that it was more like home than anything else she had seen so far in the Pacific—it was "poorer, dirtier, messier, noisier and more crowded—more like the Caribbean."

Raitea and Tahaa are the other two small Society Islands. Both are within the same large reef. The lagoon is very deep but we found a place to anchor overnight off Uteroa, the town on Raitea. The crew spent the next morning waving to dozens of small boats that passed by. It was market day.

Bora Bora

We spent almost a month here. It was good to slow down a bit and the crew really had fun with all the festivities around Bastille Day.

They happened to hit the town, Vaitape, the afternoon the festival was to start and so got a good look at the beautifully-decorated bamboo and thatch huts that housed everything--restaurants, candy stores, discos and carnival games.

Each hut seemed to be trying to outdo the others for the most beautiful decorations. Greenery and flowers on walls, inside and out, beautiful flower arrangements around bars and posts. Fruits, shells, almost everything natural except for the odd balloon or paper flower. A delightful scene Jill said. She was thoroughly charmed.

Tamure dancing and singing went on a few nights as different troupes competed against each other. Their dance is similar to my namesake, the Hawaiian *hula hula*—hip-swiveling and fast moving—but Jill says maybe a bit more "formal". The grass skirts sweep the ground. The chocolate-skinned women wear bras made of coconut shells. Their long black hair is topped with large elaborate head-dresses. Despite all that it's still erotic.

We've been anchored off the yacht club during all of this. There are several other cruising yachts including a large red catamaran with a bunch of young American hippies aboard. Frankie, the owner/captain, is planning to put on a skit at the yacht club at the end of the festival and my crew has been recruited to perform. There have been a couple of "rehearsals". Frankie will be Jolly Jack Tar—writer, director and star singer of sea chanties—and other yachties, including my crew, will be assorted pirates and bar maids. The setting is, of course, a bar. Jack has made coconut bras for the girls' bar maid costumes.

Handsome Mike has been assigned a speaking part and is dutifully rehearsing on board. Cookie will be a bar maid and Jack one of many drunken sailors. When Frankie heard that Jill was a writer he recruited her to help with the script.

***

Well! The "skit" was quite a fiasco. Jolly Jack was well lubricated even before the show started and he never stopped singing—one sea chanty after another. They were all fun but the other "actors" had a hard time inserting their lines. Mike, playing the pub owner, managed to get in a few quips but basically it was a free-for-all of singing and dancing, with the audience of yachties melding into the melee.

It was a fine send-off for Cookie and Mike, who flew away two days later, back to their real lives in St. Croix. They had been great company and we had certainly appreciated Mike as extra crew on the crossing. In retrospect J&J wondered how we would have made it without him.

We were sorry to see them go but J&J were also happy to be alone together again. And they've decided we're in the perfect place to catch their breath and get some jobs done.

The yacht club here is not really a club. It's a private business with a few rooms for rent but it's basically oriented for cruising yachts. The couple that owns it—he's German and she's French—offer everything a yachtie needs—an honor bar and small restaurant, cold showers and washing machines. 300 francs per month—a very nominal fee.

I gotta say it's about time I got some undivided attention. Jill is cleaning and touch-up painting, and she always talks to me while she's working on me so I'm enjoying the comaraderie again. I've missed that! And I'm so glad to know that Jill seems really happy and contented now. She's become a perfect cruising mate.

Jack is cutting down and remaking a mainsail passed on from Jolly Jack himself. Our current hand-me-down has about had it. It made it from St. Croix to Bora Bora without any major catastrophe but is looking quite pathetic right now. My captain has a lucky charm. Whenever he really needs something it somehow appears.

We were lucky enough to be here for a visit from Dutch Peter, a yachtie/pianist who entertained us a few nights. Then, after a quick circumnavigation of the island we took off for...

THE COOK ISLANDS
Rarotonga

Five days from Bora Bora to here, the first three light and very pleasant, the last two strong and rough.

The Cook Islands, by the way is an independently governed territory of New Zealand.

The harbor here is tiny. There's room for three small ships to tie up alongside the quay. Yachts and fishing boats drop an anchor then tie up stern-to the seawall along the shore.

There were five yachts when we came in. J&J had dinner on an old schooner one night; the schooner crew of 7 had drinks with us the next.

Then J&J embarked on the most social week they've ever had. A few evenings at The Banana Court, the local entertainment spot, where one night they saw transvestites and the next a New Zealand Maori dance troupe. They loved both!

Then a few evenings with "the princess," a jolly middle-aged woman who seems to crave the company of yachties and claims to be a descendant of the kings and queens of Rarotonga. She rides a motorbike, calls all yachties Yank, and makes every night a party, the most memorable a dinner at her house for the crews of the four visiting yachts in the harbor at the time (which were British, Norwegian and American). J&J were the only real Yanks.

Every guest contributed to a huge punch called Sailor's Suicide. The FM station's DJ, a beautiful young girl, dropped in with her ukulele made from a coconut shell, and one of the yachties had brought her guitar so the evening ended with a big sing-along. J&J loved it.

And they loved their tour of the island by motor scooter, a first for Jill. She's not happy about leaning into the curves, wishes Jack would stay more upright.

But they both loved the island, finding it wonderfully agricultural, with an extensive plain quite thoroughly cultivated. Oranges are the main crop and there's a juice canning factory. Many other fruits and vegetables. And they loved the prices, like 50 cents a pound for tomatoes.

There's a new road that circles the coast but J&J went for the lovely ancient coral road a bit inland, where all the farms are.

Palmerston Atoll

J&J had read about this tiny island and were so fascinated by the story we had to call. What a reception!

As we approached the atoll two fishing boats shot out of the lagoon, came alongside and asked to come aboard. About 10 strapping young men were our reception committee. We anchored off the wreck of a Korean fishing boat that had blocked up the small boat pass, then was dynamited so that small craft can enter between bits of the wreck.

The island is privately owned and populated solely by the descendants of one Englishman and his three Polynesian wives. How unique is that! William Marsters was just one of many seamen who went native in the early exploring days, but the only one J&J had ever heard of to do so in quite such an elaborate way.

J&J expected to find a queer, in-bred bunch of people but saw very little evidence of that. Instead they found an extremely hospitable and talented English-speaking extended family.

There are 76 Marsters living on the island at the moment, with another thousand or so in New Zealand and another bunch in Rarotongo, which they call Raro.

J&J spent an entire day ashore, visiting one house after another. There are three family groups led by three brothers and they are physically separated by "boundaries" - double rows of palm trees. J&J had to meet each one.

Bob, 67, is the policeman. He has 12 grand-children sleeping in his house.

Bill, 75, is the minister and headmaster of the school. The church is made from timbers off an old wreck. It blew away in a hurricane but was returned to its foundation.

Ned, 84, gave Jill a basket made by his grand-daughter from a coconut. She has made 400 of them and sells them for $6 apiece.

There were few members of the middle generation about, because they were all in New Zealand or Raro making money. But J&J were taken to a house where three generations of women were sitting on the floor making a variety of beautiful things. One was making a head lei out of fresh flowers. Another was crocheting. Another making a gorgeous hat out of strips of white straw. A young girl was preparing the material by scraping the topmost tender young shoots of coconut palm fronds. After being scraped very thin they are cooked and dried, then either dyed or used white.

They were all sitting on straw mats made of

pandanus leaves. Jill is in love with those and made a deal to trade one for a bottle of scotch. It is now the "rug" for our home on deck.

The next day Jack issued a general invitation to go for a sail up the coast and back. Two boatloads of Marsters came aboard including Bob and a few girls and we had a nice fast sail that everyone seemed to enjoy. J&J took Polaroid pictures for them and we think they enjoyed us as much as we enjoyed them. Jill was delighted we were really getting to meet the people.

AMERICAN SAMOA
Pago Pago (Tutuila)

Our five-day sail here was uneventful except for Jack catching a 60-pound marlin the first night out. It was his first taste of "sport fishing". Jill called it a toss-up whether he would land the fish or the fish would land him.

After all that they decided they should eat some of it, even though the fish reference book says "poor food value" and they've never heard of anybody eating one. Jill's verdict: Not great but not bad.

Jill has graduated into night navigation. She has worked out her first star sights, the second considerably better than the first, from "a rather large cocked hat to a tiny triangle." She's so surprised that after avoiding the whole concept of navigation for so long she finds she actually enjoys using the sextant, taking the sights and doing the paperwork to work out the results.

And of course it helps the night watches pass much faster. She can spend two of her three-hour watches "finding out where we are."

But don't worry, Jack is still the chief navigator. It's just good that she can pitch in.

And don't worry that she'll neglect her watch duties. She's programmed herself to check every 15 minutes—to check the sky for possible storm clouds, to scan the horizon full-circle looking for ships, to check the compass and make sure George is on course, to check the sails and make sure each is drawing correctly.

J&J are disappointed with Pago Pago (pronounced Pango by the way). After loving almost everything about French Polynesia, then enjoying the pleasant surprises in the New Zealand Islands, they're quite distressed to find this "colony" is Americanized in some of the worst ways.

Pago Pago is not the name of the island. It's the name of the capitol, which is the main town, but since it's apparently the only important place on the island it seems to be the only name anyone ever uses. Tutuila is the official name.

The town is on the perimeter of a huge natural deep harbor which is surrounded by beautiful green hills BUT one whole side of the bay is solid with tuna canning factories and attendant Korean fishing boats. The whole harbor stinks and the water is polluted.

We managed to anchor upwind of all that but it still wasn't pleasant so after a few days we moved around the corner to Fagaalu Bay, a tiny reef anchorage where a few resident boats are moored. It's quiet and calm and J&J can swim and walk or bus into town.

The town is disappointing, helter-skelter according to Jill. A sad little market, native-style general stores, big American cars and lots of beer cans and other litter in the gutters.

This is the first place in the Pacific where J&J have seen cans or any litter on roadsides or beaches.

One thing that made up for some of the disappointment was finding a laundromat—their first in Polynesia. So Jill is happily washing all the big stuff she can't wash in a bucket on deck.

This is a small island and J&J are good tourists, checking out most of it. Most of the villages are along the shore line and there are lots of small, home-made busses going back and forth.

Jill is enchanted with *fales* (pronounced follies)—the traditional houses all hand-made entirely of natural materials. There are no walls. A round or oval thatched roof (mostly coconut palm fronds) is held up by a series of posts (mostly coconut palm trunks). Straw mats, mostly of pandanus leaves, can be pulled down from the roof if a "wall" is required and more mats are unrolled for seating or sleeping purposes. Otherwise there is no furniture. This being an American island, some of the fales now have cement floors and posts and shingle roofs and are used as "guest houses".

Jill was excited that we were in the setting for W. Somerset Maugham's famous short story *Rain* and to come across the Sadie Thompson Inn. But she was really surprised to learn that the author had based his story on a real prostitute and quite upset that he had used her real name in his story!

APIA (Western Samoa)
Upolu

We are in Upolu, the capital of Apia, the main island of Western Samoa. This is a big city, by Melanesian standards, fairly attractive as it's laid out around a nice big harbor which, despite lots of shipping coming and going, J&J like. They claim it's clean enough to have a swim.

Western Samoa's most recent colonizers were Germany and New Zealand but it's been independent since 1962.

The market is exciting, lots of nice fruits and vegetables, quite cheap.

We seem to be keeping pace with the same four or five yachts that we saw in Bora Bora, Raratonga and Apia. J&J are getting to know them well and are especially interested in the crew of *Marta*, the old schooner we met in Raro, which is crewed by a bunch of young people who restored the boat in St. Thomas.

The Samoan men here are historically known for tattoos all over their bodies, including their faces, but J&J have seen just a few extreme dye cases and those were just from the waist to the knees. The men and women here dress more or less alike. *Lava-lavas* (sarongs) with shirts. Jill thinks the people both here and in Pago look sloppy and dirty.

But she is even more in love with the *fales*. There are many more here and it appears that most of the people live in them, especially in the country.

J&J took a bus tour inland and liked everything they saw. Very lush and fertile. Many villages of many people. Many more *fales* than modern houses. According to the bus driver each family has four *fales*, one for meeting, one for sleeping, one for eating and one for cooking.

Each extended family has its own chief. Over these are village chiefs, who seem quite capable of governing and maintaining law and order. The villagers are self-sufficient, growing coconuts, bananas, breadfruit, taro, cocoa, coffee, chickens, pigs and fruit and making their houses of tree trunks and leaves. The only money they need is for clothes and church.

TONGA
Vava'u

We're in Tonga's capital, Neiafu, on the island of Vava'u. We crossed the dateline getting here so we lost one day.

Our arrival was like a beautiful movie scene. As we sailed into Neiafu Harbor we were surrounded by hills with plenty of greenery and one gorgeous vista after another opening around each bend.

The harbor is a perfect hurricane hole and a beautiful one, and J&J immediately thought why don't we stay right here for the season. But the cyclone season is two months away so we will keep going. New Zealand is still the goal.

Meanwhile, they're mostly loving Tonga. Some highlights:

An anchorage off the Port of Refuge Hotel, which is currently suffering a shortage of guests so is renting showers to yachties for a dollar a head. A rather expensive shower, says Jack, but when Jill took their laundry with them and washed it in the bathtub they felt they got their money's worth.

The crafts are beautiful and J&J have splurged on a lot—all useful Jill points out. A couple of straw baskets, a tray, and the best find of all—tapa cloth. Jill has already covered a notebook with one, which will be The Pacific log.

You ought to know that since Jill started keeping a log—that's what ship's officers call the records they keep and that's what Jill calls her diary, or journal, ever since she's become a nautical person—she keeps her logs in loose-leaf notebooks, the 10"x11" three-ring hard cover ones. She types her logs on plain white paper, glues in photos or post cards where appropriate, punches three holes with her puncher made for that purpose and slips each page into the book.

Her Caribbean log cover is pasted with a few beautiful beach scenes. Her Pacific log is just getting started and up to now had nothing on the cover. Now she has glued a beautiful piece of tapa cloth to the plastic cover.

Polynesians make tapa cloth from the bark of Mulberry trees. They soak it and pound it and decorate it with beautiful designs, mostly geometrical. The colors are mostly brown and black and ivory.

Jill has been looking for these ever since we've been in the Pacific and has seen very little. Now she is happily seeing it everywhere AND has a lovely distinctive notebook.

Meeting the People

J&J had heard that Tongans were hustlers and are finding out that's true, but as Jill says they couldn't be nicer about it, and they have such beautiful things to sell!

As we left the town and visited some of the other Tongan islands we found that they are generous about giving away lots of good things to *palangis* (foreigners) but expect the hospitality to be returned, mostly in booze.

But J&J were eager to meet the people so had a few memorable encounters.

In Neiafu they met a gregarious man named Mototo who rounded up a bunch of yachties to join him one Sunday for a feast on his island, Pangimotu. We all sailed around and the crews met ashore, where the villagers prepared a feast of fish, pork, lamb and lobster, taro, yams, pawpaws (papayas) and watermelon, all cooked in coconut milk (*lolo*) and onions in their underground "ovens", all delicious.

It was followed by a show of "dancing" girls wearing leis and grass skirts... but they sat on the ground the whole time and moved only their arms! What kind of dancing is that?

Walking back from another village market one day J&J met two boys on horseback (everyone rides horses here, it's the chief form of transportation) who gave them a sack full of coconuts and pawpaws. The brothers, Vailala and Soane, spent the afternoon on board, having lunch and tea, going sailing in *Lilo* with Jack, fishing with him ashore, and making jewelry for Jill from pretty little pearl shells they found ashore.

The younger boy, Vailala, speaks pretty good English and helped Jack expand his Tongan-English dictionary considerably. I may not have mentioned that Jack tries to learn at least hello, please and thank you in whatever language he encounters. Here he's actually compiling a little list.

On another little island, Jack took *Kimo* ashore to fiberglass the bottom and met Tautai, who invited J&J for a meal on the beach that Sunday. They went and had another delicious meal baked underground but the host did not eat with them. He said that is not the Polynesian way.

They asked what they could do for him and he said not money. Jack figured he wanted booze and invited him aboard. He started on rum. He was later joined by two brothers, who quite quickly dispatched a bottle of Scotch. No one misbehaved but they were very drunk and Jack finally said it was time to go and they went. They sent back another brother with presents—a shell and bead necklace for Jill and a bead necklace for Jack.

Time to move again so next morning we sailed along the reef for a while but we were moving so sluggishly Jack decided to stop and scrub my bottom. As we were looking for a good spot to anchor an old man rowed out and led us to a nice protected bay. Jack dropped the anchor and invited our guide aboard for coffee. He turned out to be the chief of Ofu, the small island off which we were moored.

He's George, a fisherman, 73, and still going strong. He told us he had caught a huge grouper the night before and was just returning from Neiafu, where he had sold it in the market. He had rowed his tiny dingy both ways, a very long round trip.

He invited J&J to come ashore the next day for Sunday dinner. There he showed them that he was a man who had been around. Although they sat on mats on the ground outside his cabin and his *umi* had much the same ingredients as the others they had sampled his was not eaten out of banana leaves with their hands but was served on a tablecloth and eaten on china plates with silverware!

Another nice meeting, also on Ofu, was with a man who approached them as they walked along the beach and asked if Jack might be able to fix his "radio no speak". That kind of request is quite common as the natives are rarely very technically savvy but they've learned that yachties usually are.

He invited J&J into his tiny hut, which already held a family of at least 10. Jack tinkered with the radio but was sorry to say he didn't know how to fix it. They were amazingly grateful that he tried and J&J left loaded with many beautiful shells.

Before we moved on from that sweet little island J&J issued a general invitation for anyone interested to come aboard one afternoon. About 50 people came! The teacher who had instigated the visit and her friend stayed for coffee and banana bread and returned later with loads of shells and a huge shell necklace for Jill.

Back in Neiafu J&J hung out with some of the other yachties and they all had a big night out on the town. Those of us planning to go to New Zealand are passing on cans of meat to those not going there. NZ won't let you in with meat from anywhere else.

The town here is a bit more western. The houses or shacks, no matter how poor, are tidy and have tidy little lawns around them and usually magnificent flowering bushes or palmy type shrubs surrounding their property.

The people are neat and clean. The women are always in floor-length dresses or lava-lavas with tunic-type tops. The men wear trousers or lava-lavas below the knee. For dress-up—church or shopping or working in an office—both sexes wear something odd like a straw mat around their middles, tied with a beaded belt or something else fancy. They look terribly hot and bulky.

The church is powerful here. It's against the law to swim on Sunday. You need a permit to buy beer or booze and are allowed only so many bottles a month. If found drunk and disorderly on Sunday you go to jail. *Palangis* (foreigners) are not exempt. A group diving a cave one Sunday had to go to court.

The last thing Jack did before leaving Polynesia was to make our own coconut grater. Jill is really getting into native cookery. They had seen the local version of a grater ashore, where everyone cooks everything in *lolo*, coconut milk, and decided they loved the results so much they had to have the tool to make their own.

The local women squat on a low stool attached to a rounded piece of metal with serrated edges, which grates the white coconut meat into shreds. Water is added, and the mixture is squeezed through a handkerchief or cheesecloth into a bowl.

Jack made a yachtie version for us in teak and stainless steel and Jill's already cooked fish, cassava, papaya, greens and corned beef in the milk. They say everything tastes better in *lolo*.

Jack, by the way, has finally mastered a technique that never fails to crack a coconut.

MELANESIA

FIJI

Suva, Viti Levu

A four day trip getting here, going slowly with just a jib most of the time, trying to time our arrival near islands and reefs for the daylight hours. Really hazardous waters around here. J&J personally know of two boats wrecked in the last couple of weeks. One is a Wharram cat just like me which was sailing along when all its beams broke. Yikes!

We crossed the dateline again, which was confusing and didn't make navigation any easier but Jack finally figured it out. The line makes a jog east between Fiji and Tonga so we crossed it on the southern part of the jog near Tonga and the northern part of the jog near Fiji.

We're anchored off the yacht club, which J&J are enjoying. It's a big, sociable club but the local members don't mix much with the cruisers. The locals mostly stay in the bar, the yachties in the lounge.

There are lots of other yachts anchored here, lots we've never seen before, some we haven't seen since the Marquesas, and some that we've been more or less keeping pace with lately.

We're in Melanesia now and there is a definite difference in the people. The Melanesians are darker in color than the Polynesians, almost black. Their hair is kinkier and both men and women wear it in Afros, and apparently always have.

They are generally taller and thinner than the Polynesians and almost everyone wears sarongs (called *sulus* here). The women wear them as long skirts with short-sleeved tunics of varying lengths on top. They look very elegant. In town the men wear sulus made of suit material to below the knees and always a shirt.

What's left of the Wharram that fell apart trying to get here is now here. Most of the pieces have washed over a reef just a few miles from here so the owner, Dick, a New Zealander, is now putting the pieces back together again. Jack is helping him, of course.

So Jill is having fun going into town almost every day. Suva is a big city with just about everything. Lots of banks, shops of all kinds, a huge farmers' market with good prices.

She was able to have a really good long phone conversation with her parents. She has promised to call them every couple of weeks but ever since we've been in the Pacific she hasn't made a decent connection until now. The post office here actually has a special room for long distance calls, there wasn't a long wait, the operator was efficient and the reception was good

J&J are having an active social life with the big group of yachties here. They've had showers and drinks almost every night at the yacht club, barbecue dinners, Chinese dinners and fish and chips dinners. There's been back and forth visiting with several of the yachts, most of whom will also be going to New Zealand for the cyclone season.

But before we go there's a lot of Fiji to explore. It's a vast nation, with two large islands and over 300 smaller ones, only 110 of them inhabited, scattered over about 1000 kilometers of ocean.

It's a thriving country. It's been independent since 1970. Its population is half Melanesian and half Indian. The Indians were imported in colonial days to work on the plantations and many of them stayed and have prospered as farmers, shop keepers, taxi drivers, whatever. Melanesians own most of the land and most of the businesses and hold maybe half of the government seats.

Agriculture is a big thing. There are still some plantations from colonial days but also lots of smaller farms growing sugar cane and cotton and other crops.

We were told about a river trip that we could navigate from Suva on the south coast to the east coast of Viti Levu and that sounded like an interesting way to leave the big island. Can you believe we got lost?

We arrived at a bridge that was much too low for us to pass under. Jack knew that yachts with taller masts than mine had made this trip so obviously we had made a wrong turn along the way.

You should have heard Jill! We can cross thousands of miles of ocean without a navigational error but we failed to make a five-hour river trip properly!

An Indian family with a big farm along the river told us where we had missed a turn but insisted J&J visit them before we turned around.

Jack was shown around the farm by the owner and his five sons and got to watch a team of two bullocks plow some land.

Jill was expected to stay in the cook house with

the women—the matriarch and three daughters-in-law. She watched one of them nurse a huge 2-year-old, then make roti over a wood fire. She was invited to eat with Papa and Jack, while the women served them but did not join them. The menu was the roti with a hot curry sauce, rice and canned mackerel, all eaten with the hands, and tea.

Almost all the Indian women here, certainly the older ones, wear saris (Jill says they spell them sarees here), even on the farm.

We retraced our course on the river, found the junction where we had made a wrong turn, made a correct turn and continued. It was very narrow in places and shallow enough that we went aground a few times, despite my three-foot draft! We caught a group of women washing their clothes and themselves in the water. There are villages along both sides of the river and all along people came out to wave at us. As J&J waved back Jack said It's like we're the king and queen!

Back into the ocean, we started our tour of other Fijian islands.

Some highlights on the eastern side:

Levuka on the island of Ovalau. This was the capital of Fiji in early British Colonial days, before it was moved to Suva in 1882, and Jack says it feels as if nothing has changed. The Royal Hotel is full of talky ex-pats drinking on the patio. Straight out of

Somerset Maugham! Jill says. Shopping along the waterfront they found a second-hand clothing shop where Jill bought five shirts and a denim jacket and Jack bought a shirt, all for $10.

In case I haven't mentioned: Ever since we've been cruising—in both the Caribbean and Pacific—we've rarely had to change any money. The good old American dollar is welcome almost anywhere!

Kava and the sevu-sevu ceremony. Kava is an herb, reputed to be somewhat narcotic, and is a popular drink all over the South Pacific. In Fji it's called *yaqona* and has great cultural importance. It is central to the sevu-sevu ceremony in which the village chief officially welcomes visitors. J&J attended their first ceremony on the island of Nairi.

The custom is that the visitors bring a gift of kava (half a kilo bought in a village shop) to the chief. The chief welcomes them with a sevu-sevu ceremony. Cohorts mix the ground kava roots with water and serve it in a large ceremonial kava bowl (a four-legged wooden bowl). Everyone sits in a circle on the ground. Servings are then scooped up in a coconut shell which a villager passes first to the chief then each visitor and other attendants. Each person is expected to drink a cupful.

Jack drank his cupful like the locals. Jill was given a small serving for which she was grateful and managed to down it without too much grimacing. Neither of them felt much of anything. Jill had a slight reaction, numbness in her lips and tongue, as if she had had Novocain.

A leper colony. J&J had no idea we'd be visiting a former leper colony when we entered Makogai's beautiful harbor and they realized there was no sign of life in the village ashore—not a person, a dog or a boat.

They finally saw one man ashore so quickly rowed in and met Eremodo, the engineer for the island and the only one living in the colony. There's a small village around the corner peopled by other government employees. The island will soon be re-purposed as a sheep farm.

Eremodo showed them around the colony, which was built and run by the British. Lepers from all over the Pacific lived and died there. There were hospitals, houses, theaters, schools and churches. It was abandoned when a cure for leprosy was found—around 1969.

He gave us breadfruit and coconuts and came aboard for tea.

J&J also took a two-mile walk to the village, where they met the school teacher-cum scout leader, Timoci. He invited them for lunch—fried corned beef with egg and cold breadfruit eaten like bread. There are only nine children on the island now.

On Sunday Timoci came for tea with four of his students, all beautifully behaved. He brought soursop, eggplant and Chinese cabbage from his garden and when we sailed away the next day we had to stop for more gifts—cassava, part of a roasted wild goat and a sea horse his son had carved in wood for Jill!

As we sailed away Jill moaned: all we gave him was old magazines and paperback books!

J&J are loving the friendly people and the fact that many of them speak such good English they can have real conversations. Re the people they met on Mokagai Jill had to comment on their great big beautiful teeth: She said I can't help but remember that the early Fijians might have been cannibals! Depending on what history you read, that may or may not be true.

***

From there we sailed along the northern coast of Viti Levu (known as Captain Bligh Waters) for a few days, to check out some highly recommended spots on the west coast of the big island and the outer islands—the Yasawa Group in Nandi Waters.

J&J were sort of disappointed with most of those because they were quite touristy. It's good cruising ground with several nice bays but full of mini cruise ships that take three-day trips out of Latoka on the west coast of the big island.

After their wonderful experiences with the locals on the other side, this was all a let-down.

But they have learned a lot about this island nation, now their favorite. Here are some takeaways.

The outer islands are peopled strictly by Fijians in the old villages, where they practice subsistence farming. There is usually a communal garden where they grow cassava, yams and taro. They live on those vegetables and any shell fish they find and now and then a wild pig or goat. Chickens are a big delicacy even though they're scrawny.

The Indians live on the larger islands and are the shop keepers, clerks and farmers. The whole western side of Viti Levu is still covered with sugar cane, now broken up into small farms. The Indians started out as slaves on the sugar plantations and most of them stayed on and continued to be sugar farmers but many get out of the fields to drive a taxi, work in a bank or own a store.

Most of the present generations have probably never been to India but they live as if they never left there—in their dress, their food and religions. And their toilets Jill is quick to add. She is not enamored of their hole in the ground you squat over with a can of water instead of toilet paper.

Indians are about half of the population but

Fijians own most of the land. Government employees seem about equally divided but Fijians have the top positions. There is little mingling and probably no intermarriage.

J&J say Fiji is the most progressive small independent island nation they've seen in either the Pacific or Caribbean. Striving to be as self-sufficient as possible, they grow or manufacture as much as they can. They're building a huge dam in the interior so that they can have hydro-electric power.

Passage
Fiji to New Zealand

This was a 15-day passage. Not a great passage as we had a cold south wind on the nose all the way—not my favorite thing as you can probably imagine (think two cold noses ploughing directly into the waves!) We had only two squally rough nights, otherwise it was never terribly rough.

We made slow progress, tacking our way down the latitudes, everyone wearing many layers of clothes. Thanks to good old George, the self-steering vane, no one had to steer until the last day.

We had a volunteer crew aboard—a young American man named Dan who is bicycling his way around the world (and hitching rides on yachts between land masses.) He didn't start out well. He had a new ailment every day, the scariest that he couldn't pee. But luckily Jill had an antibiotic that fixed that and from then on he was fine, always willing to help, though there was little for him to do except stand his night watch and do the dishes, which he did cheerfully three times a day.

Dan is from New York State and started out on his circumnavigation 15 months ago, pedaling across the U.S. and Mexico. When he's on the road he apparently lives on peanut butter and jelly sandwiches and soft drinks.

NEW ZEALAND

I've spent the whole six months of the South Pacific's cyclone season (mid November to mid May) moored in Matauwhi Bay, part of the Bay of Islands off the north coast of the North Island of New Zealand.

It's the first time in my life I've been idle for so long. I gotta say I was ready for a rest. I mean Panama to Auckland is 6480 nautical miles! That's as the crow flies of course and with our zig-zagging from one island nation to another we surely added another thousand!

Of course we did a lot of stop and go en route so I did get some days off but basically I would say I had spent a good eight months working my hulls off! We left Tortola in the middle of March and arrived in NZ in the middle of November. That's a long work span.

So I've been lolling around in this peaceful bay where the air and water are pleasantly warm (it's been summer/fall here). I've had Jill's company most of the time, except for the month she spent in the States with her family. She's touched up all my paint and varnish topsides and below decks and made new curtains for the bunks. The ones we had were so sheer the inquisitive kids in canoes often had a good look in! We have many more islands to visit so hope to have a bit more privacy.

Jill is one of the few women yachties not working ashore for the season. She's a bit embarrassed by that, especially since their jobs are pretty grim, like frying fish and chips or cleaning hotel rooms!

But since Jack has been working almost since the moment we arrived they agree there's no need for her to get a job. They are both happy for her to work on me and get back to her writing, even though it's nothing she can sell.

The Islands column she was writing for the syndicate in the States ended when we left the

Caribbean. They figured their small-town weekly readers were not likely to be interested in South Pacific vacations.

I can't tell you what she is writing now, except for her log. It's that novel or whatever she was working on in Tortola. She doesn't talk about it much because apparently the story changes every time she sits down at that cute little Remington so whatever you might hear today is likely to be deleted tomorrow.

She loves the little summer resort town here—Russell. It's actually the biggest town in the Bay of Islands so was full of tourists during the holiday season—mid-December to the end of January.

Historically, the explorer Captain Cook anchored nearby and had a bloody altercation with the native Maoris. The oldest church in New Zealand is located in Russell and its nice old graveyard is full of Cook's sailors and the Maoris who were killed then. There is also a Captain Cook Museum with a large replica of his ship, *Endeavour*.

Jack was working ashore most of the time. First in a boat yard, getting tour boats ready for the summer holiday season, then as a motor mechanic in the town's one garage. Who knew he had that talent? Not only did he work on other people's cars he also worked on his own!

Yes, J&J toured the country by car! Jack had

found an old Trekka, the NZ version of a small Land Rover, which was actually a composite of three Trekkas, and, thanks to the tools and advice available at his job he made it work. Jill insisted on painting the eyesore, which came with a blue body, a yellow hood and orange doors. She made it all blue.

As you can imagine I wasn't happy about them leaving me for a long trip. But it turned out they covered this small country in three weeks and when they came back they concentrated on getting me ready for sea again so I forgive them.

They love the country, mostly the miles of rolling farmlands and the millions of fluffy white sheep—one million to every one person some wag told them—and the "shepherds" who are more like cowboys, riding horses or motorcycles. Of course there are still lots of dogs herding the sheep too. They range from the traditional Border Collies to every kind of mutt.

The South Island was so unique they uncharacteristically did touristy things-- a boat trip in a fjord and a helicopter ride over a glacier.

The Trekka did a good job, never breaking down. They slept in it a few nights but mostly found cheap cabins.

They were very sad to see that most of the Maoris, except for those in the tourist business, were down-trodden, drink-sodden, second-class citizens.

And they were especially forlorn to find that many other Polynesians who had immigrated here were in similar condition—such a dive from the proud happy people they had met on their own islands. Apparently Auckland—the big city—has the largest Polynesian population in the world, but sadly the immigrants have not prospered here.

Jack is finally doing some sculpting! Those dear little sheep and the unusual shepherds they saw really tickled him and he took lots of photos. Now he spends quiet evenings making small clay portraits of sheep with sweet little faces and the riding "shepherds".

He found a friendly potter ashore who let him use her kiln. He peddled a bunch of his figures to a terrific gift shop in town and has made a lot of money! What a happy interlude this has been. Now that he's finally fit sculpting into his routine Jack is determined to keep it up as we go along. Both Jill and I are thrilled with that. It's wonderful to see this softer side of Jack.

J&J have had a lot of social life here. First of all since Jill was away for Christmas Jack spent the holidays in Auckland with some yachtie friends from Tortola who had moved there. That had been planned for months so Jack was armed with presents for their two little daughters. Two adorable tiny grass skirts they had bought in Tonga just for that occasion.

Matauwhi Bay is probably the most popular bay in NZ for cruisers, especially in cyclone season, so we've seen lots of friends, old and new. There's been a great reunion of Tortola buddies that we hadn't crossed paths with since leaving there. Lots of yachts that we crossed paths with all over Polynesia and some that we've never seen before. Many nationalities but mostly British or American.

Parties were frequent, mostly on each other's yachts.

When we were ready to go I was beached for a good scouring on my bottoms and two new coats of paint there.

Almost everyone was taking off in mid-May, most to Australia but several back to the islands like us.

Although Australia is at the top of Jack's must-see list he opted not to go straight to Aussie from here but to go back to Melanesia for a few more months. Those islands are so unique and interesting he wants to see as many of them as he can. We will make the passage from Melanesia directly west to Cairns in northern Australia so that we can then work our way down the continent.

I think 10 of us set off for Vanuatu around the same day. And lost sight of each other within hours.

Passage
New Zealand to Vanuatu

We had a 12-day passage from New Zealand to Vanuatu that started with huge swells from following seas, then the wind was all over the place. Jack reduced sail one at a time until we were down to just the storm jib.

By then Jill was a case. She gets the shakes when it's stormy and I'm sure she thinks we won't make it but Jack plays everything down and always keeps his cool and reassures Jill that the wind is no more than 40 knots and the waves are less than 10 feet tall.

We later learned that we were lucky, that we were just ahead of four days of gales and that two yachts behind us were badly damaged. They limped into Vila and headed straight to the boatyard.

Once we were out of the storm the passage was almost routine. Jill was back to knitting and reading and making bread on her night watch. She did have one scary sighting on the night before landfall: a big tower that was all lit up loomed out of nowhere and a bright light fell through the sky. The tower turned out to be a large tuna fishing boat. Maybe they turned on the lights when they saw us on their radar? Maybe the falling star was a flare? Do fishing boats use flares to attract tuna?

When we finally made it into Vila J&J were thrilled to be back amidst palm trees and lovely warm sunshine. Me too. We were home again. There's no question: we are tropical flowers.

VANUATU
Vila

This is another Melanesian nation. It was previously known as New Hebrides and until 1980 was a hybrid colony of both England and France. Now it's independent. Vila is the capital and the population still has lots of French and English ex-pats.

The Melanesian natives here are much darker and fiercer looking than they were in Fiji so J&J now think that Fijians have a lot of Polynesian mixed in their Melanesian blood.

We can tell from the local artifacts that the people here are much more primitive. Weird head-dresses and masks, slit drums with carved fierce heads.

But for now J&J are enjoying the rather cosmopolitan town, off which we are anchored, as just the right size—a few square blocks you can easily walk. The shops are mostly owned by Chinese or Vietnamese (both originally brought here as slaves for the coconut plantations). A few have what Jill calls quality stuff—carved ivory figurines, cloisonné vases, leather bags, embroidered blouses. She bought one of the latter.

She was also thrilled to find British and French supermarkets with some of her favorite things like French bread, pate and cheese, a vegetable market that sold European veggies as well as local stuff, and butchers selling local meats cheaply. There's a meat canning industry here.

I know that Jill said to herself never again after that storm. But I also know that after a few days in port, she loves the cruising life. I just have to keep hoping we have more port time than sea time.

Here there's quite a large white population ashore, mostly Australians. And a large percentage of the yachties are Aussies too. It seems that the Aussie cruising circuit is New Caledonia, Vanuatu, the Solomons, New Guinea and back, and the Kiwi circuit is just Fiji and back.

Vanuatu has dealt with the condominium and language conundrums by making its pidgin the official language when it became independent. It is called Bislama and it's really cute, especially when written in the local newspaper, *Tam-Tam*. J&J have fun with it.

Here are some translations: pipol (people), pati (party), bimeby (later), yumi (you and me), nambawan (very good).

What's really exciting for me is there's another Wharram cat anchored near us! Her name is *Two to Tango* and her people are Jed and Carol. They're Aussies. We'll be cruising Vanuatu together!

THE REAL MELANESIA

Our first Melanesian village, on Epi, had a KEEP OUT sign nailed to a coconut tree. Fine welcome!

At our next anchorage, Craig's Cove on Ambryn Island, we got a good sampling of this mixed up culture. J&J and the TTs (that's what they call their companions) went ashore to explore and walked into a Catholic mission village with a school and a church. They were ignored by both adults and children, who spoke only French.

Almost next door is a Presbyterian village and school where everyone speaks English. There all the people came out to meet them and try to sell everything from pumpkins to carvings.

Apparently the two European cultures occupying the island don't mix much with each other or speak the other's language. To add to the mix are the primitive natives who escaped the missionaries. They live up in the hills and live entirely different lifestyles, still practice their "heathen" religions and speak their own languages. In fact we understand that many islands which have more than one native village have more than one native language.

The yachties got their first look at the famous Melanesian fern tree carvings and slit drums, the main art forms, here. The drums are tree trunks with long slits and sometimes figures. The carvings are simply decorative primitive faces or figures carved on tree fern trunks. Some of them date back centuries. Some of them are excellent and some are crude.

We moved on to Norsup Island, which has the third largest town in Vanuatu but has only two stores—one with bread and meat, the other with wine. There's also a post office and bank, a restaurant and a French school. J&J had a pleasant visit with two of the teachers there, a young French couple with two children who have lived in Vanuatu for five years.

Many of the huts in the villages here are made of reeds and palm fronds with high pointed roofs and are quite attractive. Others are as Jack says made of whatever's available.

Next was the small island of Vau. The population here is 700 people and J&J are sure that 500 of them came out to greet us and the TTs. Two double canoes at one time! We anchored in late afternoon and were immediately surrounded by so many outrigger canoes Jill said It's like a bumper car arena!

They were full of mostly boys and young men who were laughing and screaming. Many of them climbed aboard, up the ladder or over the sides, but there was almost no communication, not even a greeting, except for one young man who spoke some French and had brought his guitar! Of course he never got a chance to play because there was so much noise and confusion.

When it started to get dark Jack let the crowd know it was time to go home for *kaikai* and they did.

The next morning there were many more canoes, mostly with older people who just hung unto my sides and looked us over while J&J ate their breakfast and decided not to stick around any longer. A few of the locals were still hanging on while Jack raised the sails and anchor but when we turned downwind and Jack pulled up the spinnaker they grinned approvingly and let go.

We parted company with *Tango* then. They were going to move on quickly and J&J had decided to move on slowly. This is a part of the world they're unlikely to see ever again and it is so different from anything they've ever known they want to soak it up.

I won't go into details about every island but here is a brief outline of the rest of Vanuatu. Each island has something distinctive about it.

Ambrym has a volcano and some interesting carvings.

Pentecost has a thrilling initiation into manhood. Young men jump from precarious heights with nothing but a vine tied around a leg to keep them from plunging to their deaths. J&J did not witness such a spectacle but were shown the cliff where the rite happens.

Malekula has Big Nambas and Small Nambas, hill tribes that wear nothing but penis sheaths. The people in the coastal villages are all Christians and dress more or less like townies. Jill calls their dresses Mother Hubbards. The missionaries are still active and run most of the churches and schools.

Wherever J&J went ashore they were treated politely and even if they didn't want to buy anything they were often offered food out of a communal pot. J&J have deduced that the natives apparently assume that visitors are as hospitable as they are. That would explain why wherever we anchored in Vanuatu we were immediately visited by so many canoes.

I can tell you they weren't at all concerned about banging into my hulls. I was dreadfully pockmarked by the canoes and their outriggers in this area. But we couldn't scold them. They were so excited to see us and came bearing bountiful gifts of glorious fruits.

Boys and young men and occasional girls swarmed

up the ladder and over the sides, showering us with coconuts, papayas, bananas, grapefruits and other fruits we had never seen before.

They seemed to assume they would be welcomed aboard so didn't hesitate to make themselves at home, settling down on the deck, touching the white skin of J&J, giggling or shrieking with laughter. There was rarely any conversation, though now and then a teacher or bright student might make a comment in French or English.

Jill always has lots of hard candies on hand for them and makes up huge quantities of Tang to pass around but once it's obvious the group has no intention of going any time soon she often disappears below.

Jack is great about accepting the invasion. He learns how to say hello and thank you in their language and scribbles down a few other words which, we have learned, will be understood nowhere but in that particular village. In other villages even on the same island they speak entirely different languages!

When the visitors linger on he goes back to work on whatever he's doing but shows them how he's sewing the sail or splicing the rope or... When he finally asks them to leave they do so. Reluctantly.

Off one little village there was one young man who became so attached to Jack he came out several times. His English was very good and he asked all kinds of questions, assuming Jack knew everything, from how he navigated me to how big is a dinosaur and how big are super tankers and how are airplanes made. It was really touching. All Jack had to give him was a National Geographic but it was obvious that would become one well-thumbed magazine.

J&J managed to get a few walks in ashore. Once they were out of the village they were in real jungle or thick bush where the locals have their gardens spread out in clumps—taro here, banana trees there, coconut palm trees everywhere.

The natives work the old coconut plantations from colonial days, collecting copra, their only cash crop except for a little cocoa.

To continue our itinerary:

Espiritu Santo, known mostly as Santo, is the largest island in the group and has quite a checkered past. Most recently it was the center of a rebellion led by one Jimmy Stevens. Jimmy and his followers, mostly French land owners, tried unsuccessfully to stop the planned transition to independence. The Coconut War, as the international press dubbed it, was apparently waged with bows and arrows, rocks and slings.

In World War II a huge American military base covered much of the island. In addition to airfields and a city of 100,000 people it had 43 movie theaters among the staggering figures. Most of today's inhabitants live in the Quonset huts left behind but nothing else useful remains. All the trucks and tanks, anything that could be wheeled away, was driven into the sea at the end of the harbor before the Americans left. It's called Million Dollar Point and the beach is littered with rusting hulks.

Ambae: We anchored in Lolowai Bay, a very pleasant surprise—a real harbor and even a small town—at least a hospital, post office, savings bank and carpentry shop where they make furniture. Although most of the natives don't have furniture—they live on mats on the dirt floor of their bamboo huts with palm frond roofs. Ambae is the biggest copra producer of the group and many of the workers live in concrete houses and have furniture.

There is a slight sophistication about the place and there are several "Europeans" living here, Europeans being any white people. J&J met a Canadian fisheries department adviser who plans to get a catamaran and a couple of teachers including a Swedish manual arts instructor.

Our passage to the next island was squally and rainy but the most eventful ever. First Jack caught a beautiful big dorado (mahi-mahi), about three feet long. He was gorgeous—almost golden—and as he thrashed about he turned iridescent in every color of the rainbow. Jill was in tears!

As if that wasn't enough we then saw two whales spouting and diving off to port and then Whoops! Another one! A sleeping whale we almost ran into but Jack tacked just in time! He was almost as long as me! (45 feet!)

Maewo: Like Ambae, this is off the usual yacht route. The inhabitants hadn't seen a yacht for a year! But we were visited by just a few people. One of the young men escorted J&J up a precarious cliff where they viewed two spectacular waterfalls from above.

Gaua in The Banks Islands: We were visited by Mr. Patterson, the "official yacht greeter" with a visitor's book!

Vanua Lava: The first time we've been visited by swimmers, not canoes. All small boys who spoke a bit of English, bearing coconuts and soursop. J&J took photos with the Polaroid camera and gave them pictures of themselves. J&J often take pictures of our visitors which, when popped out of the camera, are considered a bit of magic by most villagers.

The camera was a gift to Jill from her first editor—in Connecticut—when he found he had a budding photo-journalist working for him.

We were also visited here by two chiefs—our firsts: the big chief of the whole island and the minor chief of the village, both of whom were very pleasant and spoke English namba one! They drank coffee with J&J.

Loh Island in the Torres Group: A very small poor village with the people literally wearing rags but very sweet to J&J. Several girls accompanied them on a walk and gave them a bunch of beautiful shells. J&J had nothing on them to offer in trade but Jack had some change in his pocket and gave them several coins. In their own currency, which none of the girls had ever seen before. Jack explained what each coin was worth. They liked the one and two cent pieces (copper) better than the 10s and 20s (silver).

That was the sweet ending to our Vanuatu experience, one which nicely wrapped up J&J's appreciation for this totally different primitive world we've just been immersed in.

Now it's time to head for Australia and J&J are having a hard time adjusting to the thought of civilization.

Passage Vanuatu to Australia

We had a good 12-day passage from Vanuatu to Cairns, most of it in perfectly idyllic tradewinds, skimming along at five or six knots. I gotta say that's my favorite way to go. With the wind behind or just off to the side I hardly have to work. The sea gently pushes me along. I can relax and simply enjoy the ride! A really nice change from beating into the wind or getting slammed in storms.

I just realized what an appropriate word beating is for me in particular. I'm the one that really gets the whipping.

Jack caught a four-foot wahoo as we cleared the islands, then never put the line in again because there was some canned meat on board that had to be eaten before entering Australia. Like New Zealand the Aussies don't allow foreign meat in.

We had two overnight hitchhikers, first a small black bird who came aboard during evening star sights, spent the night on the port cabin top and took off during morning star sights. Jack was on watch and said the bird gave a couple of thank you cries and dipped his wings as he flew away. Two nights later it was a big white bird.

Jill taught Jack how to play gin rummy. When there's down time for both of them, usually around cocktail hour, they usually play Scrabble or Backgammon but they were getting tired of those games.

In case you wonder what else they do with themselves on long passages, in addition to taking sun and star shots and working out our position in the ocean, one of them is always on watch, meaning he or she is on deck, always alert to the course and the surroundings, checking the compass, fiddling with George, the self-steering vane, if necessary, checking the sails to make sure they're drawing properly, scanning the horizon— all 360 degrees of it—to see if any change in weather or any ships are approaching.

On this particular passage Jill, who usually spends all of her down time reading, was busily knitting away. That sweater/jacket she started with New Zealand wool came out of hiding again. We'll be sure to have some cool weather in Oz.

We came through the Great Barrier Reef at Grafton Pass.

Clearing customs was a breeze, but agriculture took not just our canned meat but also all our fresh food and seeds! Jill had quite a stash of seeds for her sprouters.

AUSTRALIA
Cairns

We're here! Jack's dream country. He's never been here before but he's always been drawn to the sound of Outback and Bush. I know they'll be leaving me to explore inland later but for now we'll be cruising the east coast and the Great Barrier Reef. That sounds fantastic.

We're in Cairns, which is on the northern end of the east coast. It's a nice small city, very manageable says Jill, and friendly, with a big barn of a yacht club that's very welcoming.

Our plan is to work our way down the coast, to at least Brisbane, the capitol of Queensland. It will be a good taking-off point for J&J's inland tour and also for taking off when we're ready to head back to the islands. J&J have year-long visas so I'm sure we'll be here at least that long.

But for the time being they have decided Cairns is a good place to live and work until the new year (it's now mid-October) and that both of them will work as much as they can. The kitty needs some feeding.

Jill didn't work ashore in New Zealand because she had plenty to do onboard: painting and varnishing me and sewing new cushion covers and bunk covers etc. And because frankly the jobs available were so unsavory.

But here she was able to get the perfect job right away. The Yacht Club needed a part-time fill-in worker for bar tending and/or waiting on tables at party events. Her only qualifications are that she's a quick learner and she's anchored so close by she can be on call and on the job in 15 minutes.

She's been able to work at least a few hours almost every day and says there's nothing to it. The patrons are friendly and so laid-back they're fun to be with. They're charmed to have a Yank waiting on them and love to kid her. They also tip her royally! Yay!

She tells us that the bloke ordering a tinny is a guy asking for a can of beer. And the couple coming in for tea are there for dinner. If they do want tea they ask for a cuppa. I predict there will be a lot of Aussie talk on board the next many months.

I'm sorry to say that while Jill's been having a ball Jack was slaving away at some nasty odd jobs. One night he worked from 5PM to 6AM sandblasting a ship's bottom. Then a few days cleaning a barge's oil tanks. Now he's got a steady job helping out in a fiberglass and refrigeration outfit.

When they both have time off they do some looking around. I can tell you that within a day's drive you can see one beautiful beach after another, miles and miles of sugar cane fields and, from a train, a gorge with spectacular waterfalls.

CRUISING THE GREAT BARRIER REEF

The above subtitle is misleading. I'm afraid I

can't tell you a thing about that famous reef because we never saw it. It was a huge disappointment but J&J had not done their homework. Private yachts are not allowed to anchor anywhere along the reef. Only the tourist boats that make day trips from the mainland are allowed to do so.

So we simply made our way south off the coast of the mainland and checked out some places there.

In Townsville Jack scored a job repairing nets at a fish factory for a few days but then met a young man, Chris, who had bought a wrecked trimaran for two cases of stubbies (small beer bottles) and was looking for someone to help him make it into something sailable.

That as you can probably guess was right up Jack's alley. The tri had wrecked on the beach at Horseshoe Bay on Magnetic Island and ended up in a creek. There the two guys threw away the center hull and set about making a catamaran out of the two floats. They worked a month of very long days just to patch it up enough to float.

Then Chris got a job on a trawler for the cyclone season. He hopes to make enough money to finish the job after that. He was able to pay Jack just a little for his help but Jack was OK with that. He loved the challenge.

From there we pushed on, beating two days to

Bowling Green Cape, waiting five rainy days for the wind to shift, beating our way to the next headland, Cape Upstart.

The bad weather is not surprising. We're well into cyclone season now. But we were glad to stop in Bowen, where J&J had a nice few days enjoying a yacht club offering showers and a bar.

They also liked the town. Jill said she considered this our first real Aussie town. Cairns and Townsville are quite sophisticated and could be anywhere she said while Bowen was more what she had been expecting, a bit like a Wild West town—hot, dry and dusty, with a pub on every corner, all doing a booming business and making a booming noise.

As for the people: the women all looked fine but the men! She said most of them wear bush hats (wide-brimmed felt jobs), shorts and ankle-high suede work boots like dessert boots. She added that most of them are missing at least one tooth!

Also, in case you're interested, she added that the white collar workers wear their shorts with white knee socks, as they do in New Zealand, even in cold weather. Like old British colonial types.

From there we motored though Gloucester Pass, very pretty, then anchored off Saddleback Island for lunch and a swim. J&J have hardly been in the water since we've been in Oz. The weather has not been inviting or they've been warned to stay out because of stingers. Those are poisonous jellyfish and they're all along this coast.

J&J didn't feel so bad about missing the Barrier Reef once they realized the stingers would have made it impossible to snorkel there! They now wonder how the tour boats going out there protect the tourists?

Our next stop was Grassy Island, where J&J met a scrimshander. That's a guy who makes scrimshaw, which they love. They discovered that the carvings are done in all kinds of material, not just whale teeth. Jack especially liked a piece with white scratches on black coral. Jill liked his off-cuts (the bits left from other carvings), especially one nice squiggly creature with a yellow sapphire eye. The prices were reasonable but J&J didn't buy. They need to make money, not spend it.

We're still just beating our way south from one small island to another. The sailing isn't fun but the stops are interesting for J&J. At Airlie Beach we had a cyclone alert and Jack found a creek where we could hide if necessary but the storm went out to sea.

Airlie and Shute Harbor, our next stops, are the jumping off spots for the Whitsunday Islands, very popular destinations for both sailors and resort goers.

J&J compare them to the BVIs— beautiful with lots of hilly islands quite close together, quite green with scrubby vegetation, a lot of scraggly pine trees, some larger trees but no palm trees except those planted around resorts.

But the water, besides being full of stingers, is not clear.

Here J&J saw their first Aussie yacht charter fleets, a fledgling industry by our standards says Jack. The largest fleet he heard of was 15 boats. They plan to expand. (Back in Tortola there were several charter companies when we left and their fleets were much bigger.)

Oh, sailing into Shute we caught a big fish that wasn't in any of our books. (We've been doing very well with fish since in Aussie waters, by the way.) J&J took it ashore and nobody there recognized it either. But a trio of residents—a sailmaker, a geologist and an old English salt—were happy to cook and share it so J&J had a good time with them the next few days, checking out the nearby country, which is mostly sugar farms, and checking out a large motor boat with wheels. Really.

Once there was a seaplane that got wrecked and sank. A sugar farmer dragged it out of the water, kept the bottom part and had a boat built on top of it. When he wants to go cruising he retracts the wheels. When he wants to go ashore he drops them and drives out. Right up Jack's alley!

We're still beating our way down the coast in bad weather. We're theoretically protected some by the Barrier Reef but it's so far offshore it doesn't seem to help.

Beating out of Shute the mizzen ripped just off Happy Bay. We went in there and sewed it up and stayed two days as it was blowing up to 30 all the time. There's also a strong current running. It's full moon and they're having six-foot tides.

Finally we limped along to Lindeman Island, and the next day to Thomas Island, where J&J got such a great reception—parties at the yacht club, cars to borrow, interviews for jobs, etc,—we stayed a week.

Jack learned a little about fossicking for gold. He would love to have a metal detector but they are really expensive. $450 is the cheapest he found. So instead he bought a plastic pan for $8.50 and will start out panning for gold when they finally get out in the country.

They've decided to hole up in Gladstone for a while, as it seems the likeliest place for good jobs.

GLADSTONE

I've spent seven bloody months in this muddy river! One of them all alone while J&J were off in the

States cavorting with their families and old friends. I get it that it's important to check in with families but I gotta confess I don't like being alone. Especially in a narrow river where I'm tied to a bloody piling and can't even swing, in a town with smokestacks instead of lovely island views. I won't even mention the dirty water. It's not the clear blue Pacific I can tell you that!

We're here because there are lots of good paying jobs and J&J have to make money. Also, I'm sorry to say, they had a lot of work to do on me! My bulwarks started to rot! All four of them.

So Jack stripped them all off, then sawed off the top two inches where most of the rot was, then patched up or reconstructed what was left. This took forever because he doesn't have much spare time.

He's working for a local carpenter with a big workshop in town. He's finagled a deal to work mostly at night so that he has some of the day to work on me.

Jill has a part-time job at a health food shop in town. She likes the town. It has a good library and she's into Aussie authors. It also has a great thrift shop, where she has been able to replenish both her wardrobe and Jack's, and she loves Jack the Slasher, a warehouse/supermarket where you do everything for yourself, including writing the price on the item! What a concept!

Both dinghies are banging my port side now. Jack puts the bike in his every time he goes to work because his carpentry shop is pretty far inland. They bought a two-horse outboard for Jill's dinghy because she has a long ride downstream to her job and there's a bit of current. Her Suzuki engine is known as Suzy.

There hasn't been much social life because of the busy schedules. Once in a while when Jack is home at dinner time they will dinghy down the river to the yacht club for showers, drinks and maybe the weekly $5 steak dinner.

The only people who have been aboard are a few other yachties we met elsewhere and a German guy alone on the boat tied up on the next piling. Erhart comes over for "smoko" (a coffee break). The most interesting thing about him is that he eats out of dog bowls. The most practical solution J&J have ever heard to the monohull problem of keeping food on the table!

Our Aussie dictionary has expanded considerably. I can tell you that almost everything ends in ie. A truck driver is a truckie, a motor bike rider a bikie, a deck hand a deckie. When Jack took a day off from work with the flu he had a "sickie."

***

We're almost finished with Gladstone and we're all ready to move on. I gotta say I'm pleased with my new look. The bulwarks surgery changed the shear of my outline so I'm now looking more streamlined and longer. And of course they got a new coat of paint. The original teal couldn't be matched so Jill settled for a slightly lighter shade that I like even more.

But I'm sorry to say I didn't get out of this bloody river (formally Auckland Creek) right away. J&J were finally ready to see what they came for: the bush. So I had to wait here while they made their first foray inland.

Here are some excerpts from Jill's log about that trip.

Somehow our year in Australia was almost up and we hadn't seen much of anything—hadn't even glimpsed the outback. So we treated ourselves to a rental car and went "out west". A few miles in from the coast and you're there!

We headed for the gemfields. Jack had borrowed some gear from a bloke in Gladstone with the idea we'd be digging for sapphires but we soon discovered it would require a lot of time and work we weren't willing to invest.

We met some addicts. One middle-aged couple from Sydney went up every year for a few months of fossicking, which is what amateurs do in fairly shallow ground. They spent all day digging holes in the ground, sifting the dirt, washing the stones and once every 20 buckets or so finding a gem. But since you lose 2/3 of the stone in the cutting most of what you find is worthless.

The Sapphire Fields extend over several towns, all of which look like a moonscape because people have been digging holes in every inch of turf. Those who are into it seriously have claims, which can be from a few square feet to miles. Tourists can dig anywhere in between. We paid $8 apiece for a Miner's Right and hired equipment for $3 a day—shovel and pick, water jugs, sifter, etc. Then never scratched the ground.

But it was worth it to see how the people live. In caravans or shacks, real "humpies" or "dongers" made of old scraps of tin, wood, whatever's lying around. Some of them have built on their claim site and every bit of earth around the house is excavated.

Not far from there we felt we were in the real outback. Just dirt tracks wandering through the bush, with here and there a sign of human habitation.

We spent two nights in an on-site van at the Anakie Caravan Park in Rockhampton, the beef capitol. (Their caravan is similar to our RV—recreational vehicle—except theirs doesn't have a bathroom.) In the pub next door we met Trevor, a cowboy (called jackaroo here), who works on a "small" cattle station, only 5000 acres.

In Emerald we ran into a Country Band playing at the town hall. Jack had heard of the singer, the legendary bush performer Slim Dusty, and I was thrilled to hear him sing Waltzing Matilda, a ballad by the Aussie poet A.B. (known as Banjo) Paterson I had just discovered and loved.

At the Gregory Coal Mine we saw them digging out an open pit of coal.

At the Carnarvon Gorge National Park we saw miles of dry grass dotted with gum trees and "bottle trees" or baobab. The latter really do look like bottles. And it seems they serve as bottles too. Both the bark and leaves have moisture the cattle can extract. How cool is that!

The Fairbairn Dam which has created a huge lovely lake in the middle of the desert and where a butcher bird stole a cookie out of Jack's hand.

Several coal mines, each creating an instant small suburbia for its employees: streets full of brand new ready-made houses that are delivered on wide-load trucks.

And of course kangaroos. Many dead ones on the road I'm sorry to say but at the Carnarvon campground I got my fill of live ones. They wander around perfectly tame. With all the campers feeding them they're almost like dogs, coming right up to you, letting you pet and feed them. They don't eat meat and prefer greens and bread. The joeys are adorable and almost every kangaroo has one in her pouch. The joeys that are big enough to walk around but are still nursing do so by climbing back into the pouch head first. They rarely bother to turn around, nursing upside down with legs and tail sticking out. I loved that!

Six More Months
Bundaberg

J&J are thrilled. They got their visas extended six more months. They'll be able to see more inland and we'll be leaving in May, at the end of cyclone season. Perfect!

I'm feeling pretty good too because once we got out of that bloody river I got my yucky bottoms scraped. There was a foot of weed clinging to them! And once they were cleaned and painted my topsides got a new coat too.

We're now a bit farther down the coast in Bundaberg and the yachties we've met here think I'm pretty spiffy. The women especially love my turquoise and teal combo. I do stand out among the traditional black, blue and green hulls.

Actually we're not finished with rivers yet. We're up yet another one in Bundy, the Burnett. There doesn't seem to be much navigable shoreline on this coastline so all the towns are up rivers. But this is a much nicer town, not industrial. We call it Bundy and Jill is in love with the houses. Most of them are wrapped in lovely verandas, inviting a life that is just as much outdoors at it is indoors. That's the way to go *Hula* she confided, but I don't dare let Jack know I even think about living ashore—not yet!

Jack had his own surprise for us. First he was talking about buying a second-hand diesel engine to replace the Mixmaster, because we'll be cruising near the equator where the winds are often non-existent. But now he's been talked into two small diesels, one for each hull, an ideal solution that is affordable for us if we have them sent to our next destination—New Caledonia—to avoid the steep Australian tax!

Of course that means I'll be having some serious surgery. He'll have to cut holes in my hulls to accommodate the engines and I'm not looking forward to that. But I'm now so used to him cutting me up and I know how thoughtful he is about doing it properly that it's no big deal anymore.

He must be feeling really flush after the Gladstone jobs because he's now also got a bike. He and Jill have been sharing hers all this time but since he's discovered he can transport almost everything he ever needs—including lumber and gas—by ingeniously loading a bike, he decided to get his own.

I should say make his own. He found a bike repairman with assorted old frames and every kind of part and put them all together in no time.

How do you stow two bikes on a 45-foot

Wharram Cat? By cutting the home-made one in half. Jill's bike, made in Taiwan, is small and made to fold. It fits neatly into one of my on-deck lockers. Jack made his so the front and back could be separated for similar stowing.

Now they can do errands simultaneously AND they can tour together. They're enjoying riding around town and out into the country together!

Jill found a good thrift shop and bought her "summer wardrobe"—a skirt, shorts and a shirt. She also finally found some fabric she liked to replace the starboard curtains and sheet-cum-bunk cover. It happens to be mostly green, a point Jack had to point out was too cute, especially as the port cabin curtains are appropriately red.

It's nice that they are having fun together again. That Gladstone period was a real downer!

I've been having fun too as we're tied up to a wall just behind another Wharram. The first one I've really gotten to know. Duet is a 40 footer. Her people are Aussies—Sam and Sue—and J&J have been spending a lot of time with them. They all got quite smashed the other night when they went to town to celebrate Sam's birthday.

Australians, by the way, are known as the fourth greatest consumers of beer in the world (after West Germany, East Germany and some other Eastern

European Country). Beer bellies are common. Pubs are ubiquitous.

Jill thinks they are awful, calls their appearance the Toilet School of Design as most of them are tiled—floors, walls, even outside walls sometimes—in awful bathroomy colors. She says the idea must be you just hose down the whole place at the end of the day.

The Great Sandy Straits

This is a body of water between the mainland and Fraser Island, the largest sand island in the world. It was recently saved from sand mining by becoming a national park. It has beautiful tall trees and ferns, sandstone cliffs and lovely clear creeks. Also freshwater lakes. J&J took a bus tour and raved about how beautiful and unique it is.

It's 710 square miles.

It certainly sounds nothing like the rest of Australia!

We found a couple of nice anchorages off the shore of Fraser.

Maryborough

One more town up a river. Jack was ready to skip

it—a 20 mile river—but Sam said you gotta go because you'll see the craziest Wharram ever.

Sure enough. We've been learning that Wharram builders are generally eccentric, they tweak the plans in all kinds of crazy ways, but this was surely the weirdest. This guy was building his hulls out of packing cases. He had two old lifeboats on site that he planned to flip and use as detachable cabin tops.

His story is just as weird. He's Yugoslavian. He's an oboe player but the local employment office doesn't get many requests for classical oboists so he's on the dole. With his limited income and plenty of down time why not create a boat that might sail?

The town is billed as an historic city. Jack said that anything from the last century is considered ancient in Australia. Jill loves the few nice old colonial houses. They're built on stilts (stumps in Aussie), have one storey with a high-pitched roof and again her favorite thing—a huge veranda all around, sometimes with beautiful lattice work, sometimes with wrought iron.

She had to add how ugly the new houses are. All higgledy piggledy if you know what that means. Or just plain boxes with vacant faces. Made of hideous bricks in all colors, from pale yellow to purpley black.

Jill's getting to be quite opinionated about houses!

Every town in Queensland seems to have a pub on every corner but Maryborough is apparently the champ: 21 pubs for a population of 20,000. Jill is including such important info in her logs.

Our destination from here is Brisbane, the capital of Queensland.

It's still cyclone season so we've been limping along between northers. Back in the Sandy Straits we spent some time at Stewart Island, Tin Can Bay and Inskip Point before finally making our dash across Wide Bay Bar on the ebbing tide.

The bay may be wide but the two-mile channel between sand bars is narrow and there are huge breakers on both sides as well as big swells from behind left by the last norther. This was my ride of a lifetime! Good ole Jack was good at reading the range markers and we scooted through with a thrilling push from behind.

Mooloolaba

We spent about a week in Mooloolaba sitting out another norther. This is a small river inlet and we're on a pile mooring. It's a popular spot on The Sunshine Coast, which is apparently a classier version of the Gold Coast, the area developed south of Brisbane that is full of high rises and honky tonk towns. Here the key is canals and posh houses.

Not J&J's cuppa tea but there is a zoo and koala park in nearby Buderim and that more than made up for it.

J&J rode their bikes to get there. About twelve miles of very hilly roads! They were more pooped than I've ever seen them when they got home but they loved the park so much it didn't matter.

Jill has loved koala bears from a distance forever. She adores fluffy cute animals and these look irresistible but right away she learned she could never cuddle them, maybe never even touch them. They live in trees and have big paws and claws for climbing around and the claws are definitely not cute.

But she was able to see lots of them in the trees and is more than ever crazy about them and is now in love with wallabies as well. These are like mini kangaroos and most of them had sweet little joeys in their pouches. They were fenced but she was able to feed them through the bars and claims they have no teeth, just gentle velvety mouths.

They also got their fill of gorgeous birds, including peacocks. And a camel, a pair of buffalo, deer and monkeys.

BRISBANE

We like it!

Yes it's the big city but we're anchored off the Botanical Garden and have great views of the gardens and the skyscrapers. We're in another river but it's a big one and we can anchor out and swing! Getting ashore is fairly easy; J&J tie *Kimo* up to the garden wall or ferry dock and don't have to schlurp through any mud! So far no bugs, no smells, no black soot and plenty of breeze. The city center is within walking distance.

Jill was here only a week before she left for the States. One thing she had to do before leaving was see a fabulous exhibit now circulating the world that happens to be here now! The Entombed Warriors are a collection of sculptures of ancient soldiers that was recently dug up in China and is now on display in Brisbane. There are many separate men all wearing the same uniform but each of them has a unique face. She is in awe of the talent of those artists centuries ago.

Jack and I are like old times now. He always needs a project while Jill's away and this year he's decided my cabins need enlarging. He feels too confined below.

So he's pushed the outer edges out 10 inches and in the process got rid of the old windows. We now have tinted glass, which makes it easier on the eyes when below.

When Jack goes ashore he takes his bike and rides to outlying areas scouting for a vehicle that will take them inland and a diesel engine to make me stronger.

***

Jill returned New Years Day and things have been jumping ever since. They're excited about taking off for their trip inland AND they've made a decision about an engine for me! Engines plural!

Smart Jack did some research and discovered that the small (15 horsepower) Yanmar diesel he wants would cost half the asking price if he had it sent from Japan to New Caledonia, our next port of call. SO—are you ready for this?—he will order TWO small diesel engines—one for each hull—and have them shipped there! He's managed to convince himself he's getting two for the price of one.

And I'm really hyped about the idea. I didn't like the idea of having a heavy engine in one hull—I'd be all off-balance! But a small one on each side is perfect. I know I can carry the weight and I'll be well balanced.

I guess those months of slaving away in Gladstone were worth it. How nice to see Jack excited about spending money!

And there's more good news! Instead of buying a beat-up old van for the trip, they've been offered the loan of a campervan and—Jack's dream—a guided gold-digging tour.

They've just reconnected with a Kiwi couple we met in Vanuatu, where they were cruising with some friends. They knew we were coming to Oz so had been looking for us and spotted us in the anchorage. They are experienced gold diggers, have quite an impressive collection of nuggets in their new campervan, and were soon setting out to look for more.

They have bought a new van but haven't managed to sell their old one yet so J&J can have it for the duration!.

Sometimes we lead a charmed life said Jill.

I'm really happy for them of course. All this good news! But woe! Their good news means bad news for me. I'm going to be on the hard for four months!

The plan is to leave me in Monty's Marine Park in Caboolture, near here, high and dry on the hard (that's on dry ground) while they tour. I'll have to be there anyway when they come back and Jack performs his surgery on me and they've decided that since it's cyclone season they won't have to worry about me if

I'm already tucked in the boatyard safely. I appreciate that of course, but still. I can't imagine how boring it will be up in the air, out of the water.

Caboolture

It was a short trip down the river to the boatyard. We stopped in Doboy Creek for the night and saw the weirdest Wharram yet! Her hulls look just like mine but on top of them is a two-storey house full of household furniture and appliances! OK said Jack, it's a houseboat. But protruding from the top of the house is a mast! With a boom with a furled sail! As if they expected to go to sea and actually sail that monster!

Unfortunately no one was around so J&J were unable to ask any questions.

I haven't been aground—except on a beach to have my bottoms cleaned—since I was launched in St. Croix years ago. I can't imagine what it will be like to be out of the water, on the hard, for a long time. I admit I'm dreading it.

***

I've decided not to go into details about my two

months stay alone in the boatyard because nothing interesting happened. I learned to space out while J&J were away. I mean I was able to turn off thoughts and feelings and just exist. I just sat there on my cradle and let the sun and rain beat on me and I didn't feel a thing. I couldn't be lonely or sad or afraid. I was totally turned off. But I gotta say it didn't take a minute to come back to myself. As soon as I heard Jack climbing the ladder and shouting Hey *Hula*, we're home, I was whole again.

It was fun to hear about their outback excursion which was great but no, J&J did not get rich. He got his miner's license, bought a used metal detector and shared daily expeditions with Nick, an accomplished gold-digger, panning in creeks and digging in mullock heaps (refuse from earlier mining) all day.

Nick did make one fabulous find—a whopper of a gold nugget (8.3 ounces)—but Jack himself scored nothing but rusty nails, some buttons and an old horseshoe which did not bring him any luck.

However, they got to know the bush intimately, camping off the road, bathing in streams, cooking over campfires, stopping at small towns along the way to swim in local pools (it was HOT and dusty), pick up some groceries and in one case buy a pair of small paintings by a bush artist. Jill likes to claim that she at least came back with treasure.

Jack did have some fun. On rainy days when they didn't go beeping he brought out the clay and sculpted. He came back with a bunch of bush scenes he hopes to sell here before we take off. They're wonderful depictions of people and animals they met along the way. I especially like the self-portrait of Jack with his metal detector. Jill of course wants to keep that one.

He got to work right away and although I wasn't exactly ecstatic about his sawing in my hulls it was obvious he had planned it all carefully. He excavated under J&J's bunk on the starboard side and the salon table on the port side to build in "engine beds" for the Yanmars. He made up huge long drills for boring holes through the hulls for shafts (propellers), and cut down the skegs (projections from keel to rudder) to make more room for the props.

The props and struts etcetera he had machined in Brisbane so J&J made trips into the city every couple of weeks.

That all took a few weeks but as soon as Jack felt we were tight again we were ready to get back to the islands. Good timing because their visas were about to expire.

MELANESIA
Noumea, New Caledonia

This is a big city and not a very pretty one. It's a French island but do not believe the slogan The Paris of the Pacific.

But it's tropical! And it's an island. And there are lots of cruising yachts in the harbor. We're back in our milieu! (I hope you noticed I managed to use a French word there.)

Our engines arrived almost on schedule. Jack managed to negotiate with the ship's agents in French!

Two yachts we saw a lot of in Australia came in just in time to help Jack load the engines aboard and down through the hatches to their prepared beds. I gotta say Jack did a great job on those beds. The Yanmars look very comfy and they snuggle in just right.

Then we went up the coast five miles to Baie de Maa where I was beached while Jack cut exhaust holes in my sides. He had already made the mufflers out of fiberglass, using a small fender as a mold. Jill said they looked like dinosaur eggs but by the time Jack had added tubes for intake and outtake she had re-named them daleks. (Something from the TV show Dr. Who.)

J&J both like to name things. Jack takes the cake I think for naming our two new engines Ichiban (Number One in Japanese) and Ichitwo.

Noumea is the first port we've been where J&J have met a lot of local white ex-pats. All the yachties have "a French friend". We have a few. A day after we arrived, while Jack was repairing the mainsail and Jill was doing the laundry accumulated during two weeks at sea—you can imagine what a mess the deck was—a dinghy with four men aboard came to call.

They are all multi-hull enthusiasts. One of them has a small trimaran and is a doctor. Another is a shipwright who builds Wharram cats as fishing boats for the natives. Another lives on a yacht with his Dutch wife and children and has done a little cruising.

There was a lot of visiting back and forth during the four weeks we've been here and before we left the doctor gave us a huge bag of medical samples to update our medicine chest.

We were going to take them all, with their families, for a sail once Jack got me back together but on the appointed day—after they were all aboard—the wind suddenly switched around to the west and blew about 40 knots.

There was no way we could even get out of the harbor much less go sailing but they did stay for lunch. The party ended slowly. Jack had to make numerous dinghy trips ashore through the surf. He limited each trip to one adult and one child. Everyone was soaked by the time they got ashore but they all thought it was fun! As Jill pointed out the French do have *joie de vivre*.

Not everyone on this island is joyful, however. The locals, called Kanaks here, have recently held independence demonstrations against the French and the city looks a bit like a fortress, Store windows are boarded up and doors are barred. There is lots of graffiti showing racial unrest: colonials are assassins, the land for the people, etc.

But strangely enough J&J saw no immediate evidence of it. In fact they have commented how happy the blacks seem, especially after Australia, where they were very aware of animosity among the aborigines.

Between all the socializing Jack has been slaving away rebuilding my cabins around the engines. Both the master bunk and the salon table were returned to their original positions but two inches higher. There's also been more wiring going on.

Oh! Jack managed to sell the Mixmaster alias the Monster! We're all delighted to see that go, and to get some money for it!

Jack's been really busy almost every minute we've been here. There wasn't much Jill could do while

I was in such a mess so she spent more time ashore. I can tell you there's a fruit and vegetable market every day. The supermarkets are quite nice, all with beautiful French cheese and pate sections. The bread is cheap and delicious. A baguette costs 36 francs (30 cents).

The French have a few big department stores and small boutiques, the Chinese have dozens of tiny general stores. The prices are too high. But Jill did find some towels she liked and will recover the deck pillows with them.

Both J&J managed to get a few bike rides around the city and out to the beaches. They visited an aquarium which Jill says is justly famous.

When the engine job was finally cleaned up enough to go, and the westerly weather finally let up a bit, we took off. We were delighted to be back in turquoise water, sailing by small islands lined with white sandy beaches and feathery palm trees blowing in the breeze. Amazing how we missed that!

It was still colder than we like so we didn't dawdle, eager to get north into more tropical weather. But we did check out a few great spots:

Prony Bay: A lovely big sound with many bays, providing shelter from all directions.

Yate is a lush harbor with jungly vegetation and a tiny village, a small shop and a post office where J&J

mailed some letters.

Uvea in the Loyalty Islands—A little girl who welcomed J&J stated *C'est l'ile le plus pres de paradis*: This is the island the closest to Paradise. J&J agree!

It's a perfect atoll according to Jack, who said if he were going to design an island he would borrow lots of ideas from here. A lagoon 20 miles across is surrounded by reef islets and one long (25 miles) narrow island with a white beach all along its shore and plenty of coconut trees. It supports 3000 people who live in mostly grass huts, some of them darling little rounds.

The adults are reserved but the children are not. Walking along the beach J&J were surrounded by them, all wanting to touch them, to hold hands. One girl picked a flower for Jill, then all the others had to bring a flower. She ended up with one behind each ear and carrying a big bouquet.

The Loyalties, though still French, are Melanesian reserves, meaning the local chiefs still have the real power.

J&J looked up a French teacher they had met in Noumea. Jack had caught a fish and he and Jill ended up preparing a fish barbecue for the teacher and seven other French people on the beach that day. It happened to be Bastille Day!

We're close to Vanuatu now and J&J are tempted

to go back there. It's by far the most interesting place we've ever been but the Solomons and Papua New Guinea will probably be similar and it's a long way to Guam, where we plan to spend the next cyclone season so we've gotta move on. We'll bypass Vanuatu and head straight for the Solomons.

SOLOMON ISLANDS

A lot of the cruisers we've met say the Solomons are the best and we can see why. That is if you want to get to know the people. They are very friendly and love to be with yachties.

They are as gentle and well-mannered as the ni-Vanuatu but more confident and less shy. Yet many look even more primitive. Some have strange holes in their earlobes and sometimes their noses for wearing ornaments for ceremonial dances. Some tattoo their faces as well as their bodies.

A custom that grosses Jill out is the chewing of betel nuts. Not only does it turn their teeth black but they spit the blood-red juice anywhere they happen to be.

We've seen our first topless women, though most wear a shirt or dress or at least a bra.

Everyone seems very poor. They don't even have

a copra crop. But they seem busier than the ni-Vanuatu, making baskets, diving for shells or carving.

J&J are blown away by the carvings-- the most primitive, unique and beautiful they have seen. Almost all are made of shiny black ebony inlaid with translucent mother of pearl—components we've not seen before. Exotic human and animal heads and figures adorn everything from stylized food bowls to traditional war canoe figureheads.

Most of the sculptors are confident young men aware of the carvings' value so they are way above trading. It's the only place we've been in the Pacific where they ask for cash and a lot of it. J&J had to admit some of the larger works were probably worth $200 or $300 but sadly couldn't even consider buying one. They did score a pretty food bowl shaped as a bird with fish extremities. The artist was so smitten by Jack's jeans he was willing to trade that work of art for them.

This interchange tweaked Jack's own interest in carving to the point where he found some discarded bits of the ebony and right away spent a few hours whittling a lovely little canoe. Jill is enchanted and says we've gotta make more time for this Sweetie. You don't wanna lose your touch!

***

There are many islands big and small in the Solomon group and I think we hit each one.

Guadalcanal was the only name familiar to J&J, having read about the island as the scene of a major campaign against the Japanese in World War II. It's a big island and we've been in a number of bays with wrecked fighter planes and other war mementos.

Honiara is a newish town on Guadalcanal, created by the British as the new capital of the Solomon Islands after the war. Why is a good question, as it has a lousy harbor—almost an open roadstead. It was impossible to anchor. I had to be tied stern-to a seawall, which is a frightening mass of concrete blocks. Luckily it was very calm the week we were there. Even with five knots a swell sets in.

Despite that, J&J enjoyed being in a town again. They were out of almost everything so went on a shopping spree: two supermarkets, numerous Chinese stores, two butchers, a big daily market, one drug store.

What's more they ate out twice at Japanese restaurants, enjoyed the camaraderie of the local yacht club and saw a competition of custom dancing, which was very disappointing. They don't understand why the dancing, which was so energetic—almost erotic—in French Polynesia, is becoming more and more boring and subdued as we sail farther west.

They are also sad to find the natives who live here are mostly dejected slum dwellers. A stark contrast to what seems a happy, carefree life in the outer islands.

Here are some of those we visited. Each had something special about it.

Vanikoro: It's a few dense green mountains encircled by a reef. The lagoon inside the reef is very deep and yet has coral heads. We anchored off Nama Village in 15 fathoms!

Our special friend there was Patson, an old man with a game leg who is a diver and crocodile hunter. He was going to take Jack hunting one night but it rained so he canceled. Jill was relieved.

The canoes here are larger than in Vanuatu but do not have outriggers. Some of the big ones have sails. Patson has one of these that was grounded because of leaks. Jack repaired the cracks with bog and fiberglass.

This is a real outer island where the people have nothing. A trading ship calls once a month but the people can't afford to buy much. J&J traded all kinds of things for food: a needle and thread, matches, kerosene, soap, fish hooks, onions, rock salt (for crocodile skins) and flashlight batteries in return for coconuts, papayas, pineapples, etc.

On Rendova J&J were shown around Egola Village, a settlement of one tribe with a boundary down the middle separating the United Church goers from the Seventh Day Adventists.

In the Reef Islands we anchored between two tiny islets, one of which is uninhabited and the other of which has a small village and a well. Jack asked if we could fill our water jugs there—four five-gallon jerry jugs—and a young man named Isaac immediately recruited four women for the job.

The well was on the other side of the island so J&J enjoyed a beautiful long walk across the atoll through nice shady trees on lovely white sandy paths outlined with coral rocks. They were accompanied by the four women and a small group of men and kids.

There was no such thing as a bucket at the well. One woman had brought a tea kettle and another had a large coconut shell. These two vessels very slowly scooped up enough water to fill our jugs.

Then each of the four women hoisted a jug onto her head and started the long walk back. J&J said no no we'll carry them but there was no question. This is apparently accepted as the women's role. None of the men or boys volunteered.

Kira Kira is not a good anchorage but it is the government station for the province so has a post office and we hoped a phone, as Jill hasn't been able to call home for a month.

Well, they don't have a phone as we know it but they did have a radio-telephone that had to be relayed through Honiara, the capitol, and the station master was kind enough to let her use it and stood by to prompt her every time it was her turn to say Over. God knows what her mother made of that but at least she knows that Jill is still alive.

Actually Jill got sick soon after that. Luckily we had just made it into a beautiful haven-- Mosquito Anchorage on Malaupaina in the Three Sisters group.

Jill's problem started as a small scratch on her foot that became a tropical ulcer attended by a fever and headache. Luckily they knew the antibiotics onboard would take care of it and it did.

While they were both taking it easy they realized they both felt lethargic, an unusual condition they attributed to the heat (we're near the equator now) and the new malaria pills. In the Solomons some of the mosquitoes have developed a resistance to quinine so now visitors have to take two kinds of pills and a lot of people have complained of side effects.

Tinakula: Our first active volcano! We happened to sail by at night so were treated to a spectacular fireworks display. We drifted offshore most of the night to watch it.

***

This is the first year Jill hasn't gone home to Connecticut for the Christmas holidays. Her mother suddenly decided it was selfish of her to take Jill away from Jack at that time of year and Jill immediately concurred since she hates cold winters, so we will celebrate both Christmas and New Year together for the first time!

There's a half dozen other cruising yachts we've been crossing paths with all through the Solomons and the plan is to meet for joint celebrations. We're now in Roviana Lagoon on Rendova Island with three of them. The guys have been diving all afternoon and just came back with six lobsters they will grill ashore and eat on my deck with a big salad Jill is tossing up! A unique Christmas Eve dinner.

Big parties are almost always on my deck, since I have the biggest space. Most cruising monohulls might be able to sit six people in their cockpit more or less comfortably but with my big open deck between the cabins much larger groups can dine and lime (that's socialize in the Caribbean) and I might add dance too!

Yeah. Since discos are all the rage now, at least back in the USA, Jack has christened mine the Discodeck and what they do on it the Decodance.

PAPUA NEW GUINEA

We did not call at the highlands of this country, maybe among the most famous and primitive in the world, because they weren't on our immediate path and, frankly because J&J had overdosed on meeting primitive people. They needed a break from the harmless but pushy natives who find that white people who sail around in small boats are a fascinating curiosity worthy of their attention 24-7. They needed some alone time.

So our six weeks or so in PNG were spent on the quite westernized, prosperous and modern islands of Bougainville and New Britain.

Bougainville

Bougainville is a big mountainous island known mainly for its huge Australian copper mine—BCL. Many yachties from Australia and New Zealand head here for good-paying jobs. Most get contracts for one or two years. Others stay longer.

They can live aboard their yachts in the pretty harbor where we anchored off the small town of Kieta. The mine itself is a short mini-bus ride away, up the coastline to Arawa. There the mine has its own commercial port.

J&J know some yachties who are working at the mine so got a good tour. It is one of the biggest open-pit mines in the world. What was once a high mountain is now a big hole in the ground. The immense trucks that carry the ore from the diggings to the crushers make people look like ants according to Jill who immediately envisioned a James Bond movie in which the bucket scoops up a dozen people at a time.

I don't think she's writing a thriller but....

This island whose east coast shoreline is loaded with coconut palms also has a large copra industry. The plantations have to import most of their workers from the highlands because the locals are all making big bucks at the mine.

I got my bottoms painted before we left there. The beach at the yacht club was perfect for grounding and the full moon tide went way out so Jack raced around and managed to cover both hull bottoms in one night. Jill hovered over him with the flashlight.

New Britain

I can't believe it! We're anchored next door to a volcano! We're in Rabaul, which has a reputation as one of the most beautiful harbors in the Pacific. It is, but it's completely surrounded by green mountains that are all volcanic!! And some of them are still active! Including the one in the bay next to ours, which is expected to blow sometime soon! What the hell are we doing here!

Jill is incredulous and very upset but Jack believes in believing the scientists who grade these things and he keeps pointing out that the area has been dealing with volcanoes for centuries and the authorities surely must know how to predict them by now.

He tries to reassure her by pointing out we're only in Stage 2 Alert, meaning it should be a matter of "months to weeks" before our neighbor erupts and we'll be outta here and on our way in just a couple of weeks. (Stage 3 is "weeks to days" and Stage 4 is "days to hours".)

We ended up spending three weeks here. Jack was especially interested in the amazing command center the Japanese navy had constructed when they took over the island in WWII. A fortress of caves and tunnels carved through mountains. Ships were hidden there. There was even one deep enough for submarines. The admiral's bunker is now a war museum in the center of the city.

Otherwise the town is attractive and Jill was relaxed enough to enjoy the shops and especially the market, where avocados were in season and sold for 10 cents. If you showed up after the market ladies left you helped yourself to whatever they left behind.

She reports that until recently—pre-volcano scare—the town had a big population of ex-pats, mostly Australians, about 2000 of them, but most of them have already evacuated. Also that sarongs are called lap-laps here and are worn by both men and women. The women wear them long with loose short-sleeved blouses.

Then she got another tropical ulcer. On her foot. She had been treating the scratch with antibiotic creams for weeks but it suddenly erupted—her word, obviously volcanoes on her mind. The foot got very swollen and the wound got black around the edges.

She could hardly walk so Jack found a taxi and a doctor who scraped off the black stuff. Dead tissue, he said, like gangrene. He opened a hole to let the pus out and gave her some antibiotics. It's getting better very slowly but there's still a lot of raw flesh she cleans and covers with 2x2 inch bandages. She's starting to walk on her toes.

The nice thing about this is how sweet Jack has been. He's done the shopping and even the cooking for days. He's always thoughtful and helpful but it's really obvious now and Jill is so in love all over again she's forgotten to worry about the volcano next door.

PASSAGE MELANESIA TO MICRONESIA
Rabaul to Ponape

We crossed the equator when we left Melanesia and entered Micronesia. The wind was predictably non-existent there and the westerly current was quite strong so Jack was thinking we'd never make it to our intended destination but then the trade winds picked up farther south than expected and we just shot up north.

Still the passage was a long 11 days. We were close-hauled all the way but the sea was slight to moderate so it was an easy run.

We had some good fishing, including a three-foot shark which Jack gave back to the sea and a four foot barracuda, which J&J ate some of. They have learned that unlike Caribbean barracuda, which is poisonous for humans, the Pacific variety is edible and quite delicious, though the smaller ones are better than the big ones.

Jack's main interest in Micronesia is to see the sailing canoes that are still made here by hand, using only natural materials, and hopefully to meet Mau, the navigator in the documentary of the same name that inspired him to make me!!

The islands harboring both the sailing canoes and the navigator are at the far end of what used to be called the Caroline Islands but since WWII is the American-controlled Federated States of Micronesia.

THE FEDERATED STATES OF MICRONESIA

Ponape

Micronesia means tiny islands and now we know why. Ponape is one of its biggest islands and it's only about 14 miles in diameter. It is an atoll really, entirely surrounded by reef.

Within this atoll is a sad little town and a fabulous ruin.

The USA cannot be proud of the town of Kolonia, which is the capital of FSM. It's an eyesore. Most of the buildings are Quonset huts or abandoned warehouses, all rusty. The market sells bananas and cucumbers and little else. The ground is littered with garbage. America introduced the supermarket but seems to have neglected the concept of garbage disposal. Bottles and cans and all the packaging from the American supermarket litter the ground.

J&J have met some of the Americans who live here. They say that most of the U.S. money that pours in here seems to disappear into the pockets of a few local politicians. Complaints and exposes go nowhere because everyone is related to someone he wants to protect. Just like home say J&J. The USVI has a similar complaint.

The fabulous ruin, totally unrelated to anything else in this part of the world, is Nan Madol, an ancient city built around 1000 AD. A fortress made of large basalt logs, many of which are still standing. Archeologists speculate it was built by invaders from Kosrae, 300 miles away, as their new governmental and military headquarters.

The whole complex is built on a reef and is approachable only at high tide. Canals were dug into the reef for canoes. J&J checked it out from the dinghy.

Truk

Truk may be the biggest lagoon in the world, 40 miles across. It was a major Japanese naval base in WWII and is now a scuba diving mecca thanks to the 60 something ships sunk by Americans during the war. J&J don't scuba dive but they were able to get an idea of the destruction by snorkeling.

They were happy to see the town of Moen was

more attractive than Kolonia—not great but at least there was some attempt to pick up garbage.

We anchored out of town, off the only resort hotel, which turned out to be an excellent decision because we were stuck there for three weeks when Jack came down with malaria and was quite sick!

J&J had taken malaria pills all through Melanesia, where the disease is common, but stopped when we got to Micronesia, which is free of the disease. Apparently Jack had contracted it while we were in the Solomons or New Guinea. In retrospect Jack realized he hadn't been quite up to par lately. The pills had kept it dormant until they wore off!

After a few days of chills alternating with a high fever, classic symptoms, Jill was able to get him to the local hospital, where they were able to diagnose him but had no medication for malaria because they never needed it. Luckily Jill had stashed some cure pills just in case and thank god they started working immediately. There were only three pills in the stash but they did the job.

The upside of this ordeal was Doctor Dick, a young American working for the Public Health Service at the hospital, who befriended J&J. He voluntarily made boat calls for us every evening after work for a couple of weeks by riding his motorbike several miles over a rutty, muddy dirt road, then sailing his windsurfer, which he kept at the hotel, out to us.

He was able to acquire a long-term cure for malaria, primaquin, which he provided for another two weeks. Jill's three-day cure apparently gets the bugs out of the system but doesn't kill any eggs that might be hiding in the liver and re-emerge months or years later and cause further attacks. Dick's follow-up cure is supposed to knock out everything so the disease can never recur.

Fingers crossed says Jill.

Dick is now a best friend of course and he's brought out his girlfriend Liz, a Peace Corps nurse, a few times. When they finish their terms here they've decided to go back to the States to make enough money to buy their own boat and return to the islands as mobile floating medical practitioners.

PULUWAT AND SATAWAL

We're finally here at the only islands in the world that Jack would not miss. He hopes to see how the sailing canoes are made and to meet the man who inspired him to build me.

The neat thing is that while Jack is ecstatic to finally reach two goals, the people of these two little islands seem equally thrilled to have us! Both islands are in desperate need of a major fishing expedition and as soon as they saw me they thought their prayers had been answered.

Both islands are tiny specks in the sea (about 14 square miles) with populations of maybe 600 people each. Each is at least 150 miles from an island with a town and its only connection to the rest of the world is a trading ship that arrives more or less monthly bringing rice and other food staples and mail or other messages. Otherwise they must be totally self-sufficient.

Fishing is obviously their principal source of food but the waters nearest the islands were fished out long ago. Now they must sail their canoes, less than 30 feet long and carrying a crew of six to eight men, up to 150 miles away to find a reef, bank or atoll with a decent catch.

When they saw me—45 feet long and 20 feet wide with plenty of deck and net space, they foresaw a longer sail and a much bigger catch. Jack was delighted to oblige.

We hosted a major fishing expedition for each island. The biggest was from Satawal to the uninhabited atoll of West Fayu, an overnight sail. Our crew of 10 or 12 burly men in colorful loincloths included the revered Mau but J&J were disappointed that when he took the wheel, instead of demonstrating his navigational knowledge for them, he was happy to stare at the compass.

After sailing all night, at daylight each of the men went into the lagoon with spears or nets, and by dusk every inch of my decks and nets was covered with seafood—fish in all sizes, clams and lobsters, even giant sea turtles.

Back at Satawal the next day, after another all-nighter at sea, the crew had the assistance of the entire island population in unloading and storing this very welcome bonanza.

On both islands Jack was welcomed into the men-only huts where they make their canoes and he was finally able to witness the creation of one of their proas, a canoe with one outrigger.

They make small proas that can be handled by one man but the voyaging canoes that make the long fishing trips require a crew of six or eight men, even though the canoe is only 25 to 28 feet long! Because of the one outrigger they can be sailed only upwind, which means the mast has to be moved every time they tack. The bow and stern are interchangeable.

No wonder they were so happy to see us! said Jack.

Their materials are all indigenous, all found on their island. The hulls are hollowed out of breadfruit or mahogany trees. The men chop away with tiny adzes for months and months. Planks are sewed on with rope made from coconut fiber. Caulking is breadfruit sap that sets up like cement.

Their sails used to be made of pandanus mat, so heavy that every time it rained the sail had to be lowered. They have now switched to dacron sails. That and a bit of metal in the sheet block are the only imported items they use.

Their ocean is the vast Pacific, which covers almost half of the planet's surface. Here's a figure Jack dug up: 57 million square miles! (Believe me, it's hard to believe I've already covered a major portion of that expanse in the last four years!)

Centuries ago some of these natives' forefathers managed to explore this huge territory, sailing their home-made canoes on months-long journeys to discover other islands. They often sailed against the wind and currents, relying on the sun, stars, clouds, seas, swells, birds and fish as their guides.

The navigational knowledge they acquired has been passed down orally and through drawings in the sand through generations. The few remaining navigators are middle-aged men now and, we learned, few in the next generation care about studying the traditional lore. They know there are compasses and sextants to do the work for them. And Jack is sure they will soon know about satellite navigators, the new gizmos we just learned about that the yachties are calling sat navs.

GUAM

We've been here almost a year. It was time to stop cruising for a while and make more money. J&J were glad for a change from primitive Micronesia.

We by-passed the Philippines because J&J had heard frightening stories about the crime there.

We're geographically still in Micronesia but everything is so different. The indigenous population is Chamorro, like in the Philippines, much lighter-skinned, and there are a lot of Asians. Both the Philippines and Guam were Spanish colonies for centuries, until they became U.S. properties after the Spanish-American War.

Guam is a little larger than St. Croix with a slightly larger population. J&J find it significant that it is the westernmost U.S. possession, as St. Croix is the easternmost. So we have traversed all of our country's range.

About a third of the population is American military, Air Force and Navy.

We got here in time for typhoon season. Now that we're above the equator cyclones have yet another name and the season is similar to that of hurricanes in the Atlantic.

We were anchored in Apra Harbor the whole time, off the Marianas Yacht Club. (Guam is in the Marianas group of Micronesian islands.) It's a huge harbor and we share it with the commercial port around the corner and the Navy base across the harbor but despite all that the water is fairly clean. Jill swims laps around me almost every day.

The typhoon season was quite active and we made a few trips up a creek when there were scares. Twice Jack tied me up to several trees, but we never felt more than 80 knots of wind.

The yacht club is not at all swank. It has a leaky roof and other issues. But there's a nice patio with lawn chairs and barbecues and cold showers and a washing machine and dryer. And a snack bar on weekends when a lot of residents come out to sail their small boats.

Jill was especially delighted by these comforts of home because she spent a few weeks here alone while Jack flew to Hong Kong and delivered a large motor yacht to Taiwan. I'm not sure how he got that job but he knows the other two crew members and it paid enormously so of course he went.

Otherwise he's had a steady 9 to 5 job with a construction crew. He likes his fellow workers, especially when they recognize his carpentry skills and give him the "finish" jobs. And Jill has had a part-time job at one of the many tourist hotels on a beautiful mile-long beach.

Tourism is a big thing here and most of the tourists are Japanese. That amazes Jill since Japan was a brutal conqueror of this island for a couple of years before the US won it back at the end of the war.

Apparently all is forgotten/forgiven on both sides as the locals and the tourists seem to get along fine.

Jill's job is in an office, typing up whatever the hotel needs typed. It's boring but it's only three hours in the morning and it pays extravagantly because good typists are hard to find here.

It's perfect for Jill because she's now taking art classes three afternoons a week! She's often thought she'd like to take an art class and here she has met an art professor who has invited her to audit his drawing class at the university—for free! He lives on his boat across the harbor at the Navy yard but often dinghies over to the yacht club at happy hour.

She's getting quite good! By her second semester she was drawing faces, something she was sure she could never do. Her self-portrait, homework done in pastels by sitting on the head to access the only mirror aboard, came out recognizable. And her final effort is a real keeper—a full-length portrait of a young man reclining on a lounge chair—done in the classroom with the unusual medium of a magic marker!

Now we've got two artists in the family, crowed Jack. And you've inspired me to get out my bits of wood and see what I can do with them.

Another first: Both J&J have a vehicle! Both were on their way to the junkyard when Jack grabbed them and was able to get them back in working order! He goes to work in a 15-year-old station wagon christened "the gas chamber" for its foul exhaust fumes that inundate the car, and Jill goes to her destinations in an 18-year old pick-up truck lacey with rust but considered more reliable than the other.

I'm so glad she's found a new interest because I'm getting worried about her. It's almost time to head for the Indian Ocean and I know she's really afraid.

We've been hearing nothing but bad stories about everything along our intended route. The Philippines where piracy and theft stories abound, including one about several yachts being robbed of major equipment including navigational gear and anchor chains one night! The Strait of Malacca where among several piracy reports there's one about a yacht we met in Vila. The English couple and their two children were killed and tossed overboard!

And I know that Jill worries that if we ever do

make it across the Indian Ocean to Africa there's still rounding The Cape of Good Hope. That has been her big nightmare all along. It sounds like the roughest passage in the world.

Despite all the worries she hasn't backed out. But I do know that when J&J left me up the creek for a month while they went to the States to see their families Jill went to St. Croix for a few days and can you believe it? She bought a lot! A lot to build a house on!

She had gone there to see her friends Cookie and Mike. Cookie has gone into the real estate business and Mike is now the first vice president of his bank and they're both making lots of money and willing to finance their dear friend. They knew that Jill would never dream of ruining Jack's dream by backing out of the circumnavigation but desperately needed something joyful to look forward to after completing it. A dream house in St. Croix was their solution. It could be rented out while J&J continued around the world.

Well, Jill did not find her dream house but that did not stop Cookie. They went on to look at available lots and—serendipity—found the perfect perch on a hillside overlooking Salt River Bay, where I could be moored. The price was low enough that Jill felt she could ask her parents for a loan and they gladly obliged. They are not thrilled about the Indian Ocean either.

Before we finally left Guam Jack managed to sell both junk cars, for exactly what he paid for them and a few little wood carvings of boats to a local gift shop, which paid him handsomely.

THE PHILIPPINES

We came here after all because Jack suddenly decided he shouldn't pass it by. Despite all the bad things we've heard about the Philippines I'm happy to say that so far we've had no bad experiences and J&J have mostly enjoyed all the places we've been and the people they've met.

Our first port of call was Caiman Cove in Dasol Bay on the northwest coast of Luzon, which is the big main island and the seat of the capitol, Manila, farther south.

The cove is a lovely spot with two villages ashore and friendly people. There is no road access. The nearest market is about an hour away by *banca.*

Bancas are the local boats—canoes with outriggers on both sides. Precursor of the trimaran, says Jack. The floats are bamboo poles attached to the hull by bamboo arms, making the whole thing look like a great big floating spider says Jill. Here they all have diesel engines but we hear there are also sailing *bancas*.

Everyone grows rice and uses buffaloes (which they call caribou) to haul their rough handmade carts, few of which have wheels.

There are ponies for riding, fresh water wells, a fresh-water stream for washing clothes, plenty of bananas, papayas and coconut palms but unfortunately no longer any market for copra. The houses are mostly bamboo with thatch roofs.

J&J's special friend there was Pacifico, a retired high school English teacher, now one of several store keepers in the cove. A store being a small section of a house. His has beer and coke and a few canned goods, sugar, candy and cigarettes. Everyone smokes here although everyone is poor.

All the men seem to fish every night from their *bancas*. If someone catches more than his family needs he sells the rest.

Fishermen are still allowed to use dynamite here so snorkeling was terribly sad. All the reefs are dead!

Our next stop down the coast was Masinloc, a small town with a recently made road into Manila. J&J did not want to sail into Manila Bay, where they had to do the formalities, because it is said to be filthy, it takes a day to sail in and there can be pirates en route!

It took a while but J&J were befriended by the town wholesaler, Pedro, who gave them a day-long truck ride into the city when he went for his weekly buying trip and put them up for a few days in his very basic cold-water apartment.

They got a good look at the countryside, where rice is growing everywhere and whole families spend the day working in the paddies.

But the most interesting thing was the jeepneys. After the war the U.S. Army left all their jeeps behind and the locals turned them into public transportation, sort of mini-buses. Over the years they have elongated the bodies and replaced the engines with Japanese diesels and decorated the bodies with everything imaginable. A lot of shiny chrome mixed with brilliantly painted scenes and slogans, streamers and religious symbols.

There didn't seem to be bus stops. It was more like door-to-door service. If you needed a ride you stepped out your door and a jeepney would stop to pick you up.

The other major form of transportation is the tricycle—small Japanese motor scooters with homemade sidecars. These are also highly decorated and very cheap, a matter of pennies to go miles. Like the jeepneys they are usually overcrowded. Six on a tricycle is routine.

J&J were charmed. No one seems to be in a hurry and everyone is very good-natured.

The city itself was uninteresting. Manila was almost completely bombed out during the war so almost everything is fairly new but J&J did see what was left of an old walled city, including one beautiful church. Almost everything else was grubby and dirty and poor. The pollution is terrible, blamed on many diesel engines.

At night their host Pedro gave them a tour of the nightlife hot spots including a discotheque owned by his cousin in the tourist area—one dive after another with go-go girls.

J&J were told their little tour inland must include Baguio, the summer capital, a city 6000 feet up in the mountains that was developed by the Americans when they were running the place. They took a five-hour ride in an air-conditioned bus to get there. It was certainly cooler. It's an agricultural center for non-tropical foods. The market had asparagus, cauliflower and strawberries! There were also some lovely handicrafts like bamboo baskets and mobiles. But otherwise they were disappointed.

They're glad they got to see some of the interior. They enjoyed riding around in busses and trucks, looking at the countryside and the way people live. In the country it's almost all bamboo huts with thatched roofs amid coconut palms, banana trees and papayas. In the towns and cities it's all cement buildings and shanty towns.

Back in Masinloc we had a message: A boat builder in Cebu, an island farther south in the Philippines, has a job for Jack. He needs a helper to finish an order for a catamaran and heard through the cruising coconut vine about Jack. Can we be there ASAP? Needless to say we're off.

Just what we need said Jack. A few months of good pay before taking off for the Indian Ocean.

Cebu

Jack is ashore almost all day every day as he and the boss rush to finish the catamaran that is nothing like me! It's a racer! A light, sleek racer! It's going to be in an international race in New Zealand, the first multihull to ever enter such an event. Wow!

Meanwhile Jill is in charge of getting me ready for the long crossing. Both maintenance and provisioning. She's really stocking up, so afraid of what's coming up. She wonders if they'll ever have another chance to shop. She's confiding in me like in the old days, when everything was new to her. Now she's an old hand but she's afraid again. I hate to see her so worried. She tries to keep her fear from Jack.

***

Oh No!

Jill has a lump on her breast. A great big lump just suddenly showed up! J&J are devastated!

Luckily they know a woman on one of the other boats in the harbor who just had breast cancer surgery right here so they immediately met up with her. She told Jill to see the Chinese doctor in town who had treated her and Jill went there.

She found the office dirty and the doctor scary. He took a quick look at her lump and announced you have breast cancer, come upstairs to my operating room and I will cut it out now. She got out of there fast.

Back on board she fixed herself a stiff drink and stretched out on the deck cushion, hugging all the pillows.

It's true *Hula* she cried. I've got cancer! What do I do?

By the time Jack came home she had decided.

She would go home to Connecticut and have surgery there and since there would doubtless be months of chemotherapy treatments to follow Jack was not to wait for her. In fact he should find new crew and complete the rest of the circumnavigation without her.

Wait a minute Jack said. You're upset. You're in shock. Take your time. Yes go to Connecticut for surgery but you're going to get better and I can wait for you. We're doing this together. Whenever you're ready.

Jack helped Jill pack her stuff. There were just two small duffle bags. Before Jack took her to the airport she broke down saying goodbye to me. She said I love you *Hula*. Take good care of Jack. As you can imagine I was a mess.

***

That is not quite the end of this story. When Jill saw a surgeon in Connecticut he said her lump was just a cyst. He lanced it and it went away. When she called Jack with the wonderful news he was ecstatic, assuming that meant she'd soon be back onboard.

But then Jill said no. I'm still not coming back Jack. I hope we'll be back together later but I just can't face the Indian Ocean and the rest of it. I intended to tough it out but I've been terrified of what was coming up so when I thought I had cancer I was almost glad, it almost seemed like a blessing. It gave me the excuse I thought I needed to back out. I hope you can accept that I love you and hope we'll be together again later but I just can't do what you're doing anymore Sweetie. You'll be fine without me.

Jack was speechless.

Jill said I'm sorry sweetie. I love you but that's it. Take good care of *Hula* for me. I'm leaving for St. Croix next week. I'll see you there whenever.

Epilogue

A few years have passed since Jill left. That turned out to be the end of Jack's circumnavigation dream.

When he was in Cebu after Jill left he continued working for the boat builder. A young Filipina entered the picture and a year or so later produced a daughter. Jack broke the news to Jill in a letter, casually mentioning that he was now a father.

Luckily Jill was enjoying being back in St. Croix, writing for the paper again and loving the sweet little house she had built and planting a garden and generally enjoying life. Much to her surprise that shocking news didn't destroy her. She's moved on.

We eventually moved to Hong Kong and there are two daughters now. Jack loves being a father!

We're not in the crowded city thank god but out in a suburb, anchored in a pretty little bay of clean water with a small marina and a nice yacht club ashore where our crew has guest privileges. Jack's working with boats so is in his element. We are here probably for the rest of our lives.

I think Jack's fine about giving up his dream. And so am I. It's kinda sad to ditch a dream like that but hey, we had many wonderful years of cruising. It was great while it lasted.

Meanwhile I'm so happy Jill has achieved a couple of her dreams. The house which she started planning while still aboard was built within a year. It has a long gallery much like the Australian verandahs she so admired and has a gorgeous view. On a hillside, it overlooks the Carribbean Sea and a lovely bay. Even though the bay's anchorage does not include me she is happy. She is painting and writing and has published her first book! It's about the cruising life! Jack and I both get wonderfully high marks!

What makes me really happy is that J&J are still friends. They communicate now and then and obviously still care for each other and appreciate what they had together, including me.

Me? I'm fine. I still get plenty of attention from Jack and his children are a delight. There's nothing sweeter than the giggling of two little girls.

I realize I might never go to sea again and I'm okay with that. I do miss gliding through the sea on a sparkling day but I'm content enough just lolling in this pleasant anchorage and now and then recalling the good old days with J&J.

I like that this story has a sorta happy ending.

#

Nauticalese
(A Glossary)

Aft: Toward the rear of the boat.
Below: Under the deck.
Bilge: The area inside the bottom of a hull.
Boom: The horizontal post to which the bottom of a sail is attached.
Bow: The front of a vessel.
Bulwarks: A "wall" enclosing the perimeter of a deck.
Bunk: A built-in bed.
Cabin: Interior area with standing room.
Cleat: A two-pronged metal base on which a sheet is secured.
Companionway: Ladder or stairs from deck to below.
Crossbeams: Beams connecting the two hulls.
Deck: The horizontal part of a ship.
Dinghy: Small boat that takes you ashore.
Dodger: Canvas protection on deck.
Fore or Forward: The front part of a boat.
Galley: Kitchen.
Genoa: A large jib sail.
Hatch: A horizontal hinged "door".
Head: The bathroom and the toilet within.
Halyard: Line holding sail to mast.
Hull: The shell of a boat.

Jib: A small sail used forward of the mainmast.
Lifeline: A plastic-covered wire that circles the boat.
Line: Rope.
Mainsail: A large sail for the main mast.
Mast: A spar that carries a sail.
Mizzen: Both a smaller mast aft of the main and the sail on it.
Port: The left side of a boat facing forward.
Porthole: A small window in the side of a boat.
Reaching: Point of sail when the wind is on the side.
Rigging: The lines that connect the sails to the masts.
Rudder: A steering slab off the rear of a boat.
Salon or Saloon: The living area of a boat.
Schooner: A boat with two or more tall masts.
Starboard: The right side of the boat facing forward.
Sheet: The line on the bottom of a sail that winds around a cleat.

Made in the USA
Middletown, DE
05 November 2023